DC Wind Generation Systems

Omid Beik • Ahmad S. Al-Adsani

DC Wind Generation Systems

Design, Analysis, and Multiphase Turbine Technology

 Springer

Omid Beik
Department of Electrical and Computer Engineering
McMaster University
Hamilton, ON, Canada

Ahmad S. Al-Adsani
Department of Electrical Engineering Technology
College of Technological Studies (CTS), Public Authority for Applied Education and Training (PAAET)
Kuwait City, Kuwait

ISBN 978-3-030-39348-9   ISBN 978-3-030-39346-5   (eBook)
https://doi.org/10.1007/978-3-030-39346-5

This Springer imprint is published by the registered company Springer Nature Switzerland AG
The registered company address is: Gewerbestrasse 11, 6330 Cham, Switzerland

Preface

Wind energy systems have been the center of much attention in recent years, resulting in many research publications and an increasing number of installed systems by industrial manufacturers, electrical grid operators, and utilities. This book discusses existing AC systems, all DC wind generation systems, their components and associated technologies. The DC system uses medium voltage DC (MVDC) array grid, DC/DC offshore substation, and high voltage DC (HVDC) transmission. Unlike the conventional schemes, the turbine conversion system incorporates concept of multiphase high voltage hybrid PM generator and passive rectification stage. Mathematical models for the wind turbine operation are extracted and verified via comparisons with published commercial wind turbine characteristics. Two regions of operation for the wind turbine are proposed: a limited speed region and a high speed operation region. The operating principles for limited and high-speed operation are explored to study its operational implication on the HV hybrid generator design. The detailed machine design study focuses on the limited speed region which is the current industry practice for wind turbine control and operation.

The DC wind generation scheme is compared with commercial AC offshore wind systems to identify the system envelope specification requirements, key technology gaps, benefits, and system performances and deduce component ratings. A feature of DC systems is the reduction in power conversion plant at the turbine, a feature that reduces cost and improves system reliability. Specifically, the active power electronic converters and turbine transformer used in existing commercial systems are eliminated in the DC system; therefore, system complexity and component count are reduced. The DC system offers a variable controlled voltage to the local wind farm interconnection and a fixed voltage for transmission through variable step-ratio DC/DC converters. The variable voltage feature is of importance as it allows for greater controllability, a feature that is not feasible for the existing AC wind systems. The book in its final chapter discusses insulation system design for the high voltage generators based on commercial high voltage winding practice with the conclusion that the winding scheme is highly practicable.

The author wishes to extend special thanks to Dr. Nigel Schofield at the University of Huddersfield for his valuable inputs and also to the team at Springer for their care during the book production.

Toronto, ON, Canada

Omid Beik, PhD

Contents

Chapter 1
Wind Energy Systems

1.1 Introduction

Wind energy systems have been the center of much attention in recent years resulting in many research publications and an increasing number of installed systems by industrial manufacturers, electrical grid operators, and utilities. Wind energy is a renewable and sustainable source of energy that is environmentally clean with regard to CO_2 production. However, wind energy varies with the wind velocity; therefore, it needs to be appropriately harvested to be useful. Wind turbines are used to capture the kinetic wind energy and transform it to mechanical energy which is then converted to electrical energy by electric generators. The flowchart in Fig. 1.1a shows a conversion scheme from wind to electrical energy. Basic power conversion train for a wind turbine is presented in Fig. 1.1b. Wind turbine blades capture the energy in the wind by slowing down wind velocity. This energy is then transferred to the turbine rotor which spins a generator, and electrical energy is produced. A gearbox is sometimes used to increase the turbine rotor speed to higher speed suitable for the generator. If no gearbox is used, the turbine generator is called direct drive generator (DDG). Existing turbine power trains use industrially rated generator technologies such as the induction generator (IG), synchronous generator (SG), doubly-fed induction generator (DFIG), and permanent magnet (PM) generator either with or without a gearbox. The turbine hub, shaft, and other rotating parts are referred to as wind turbine rotor. Figure 1.1c shows a 2.7 MW commercial wind turbine, ECO 122-2.7 MW, manufactured by Alstom. The turbine conversion power train is located at the top of the turbine tower, i.e., in the nacelle. Approximate length, size, and mass of main turbine structure for the Alstom ECO 122-2.7 MW are shown in Fig. 1.1c.

Modern wind turbines normally have three blades with the capability of changing the angle of the blades with respect to the plane of the rotation of the turbine. This angle is called the pitch angle and is an important parameter in wind turbine control. Figure 1.2 shows main components inside an Alstom wind turbine.

© Springer Nature Switzerland AG 2020

O. Beik, A. S. Al-Adsani, *DC Wind Generation Systems*,

https://doi.org/10.1007/978-3-030-39346-5_1

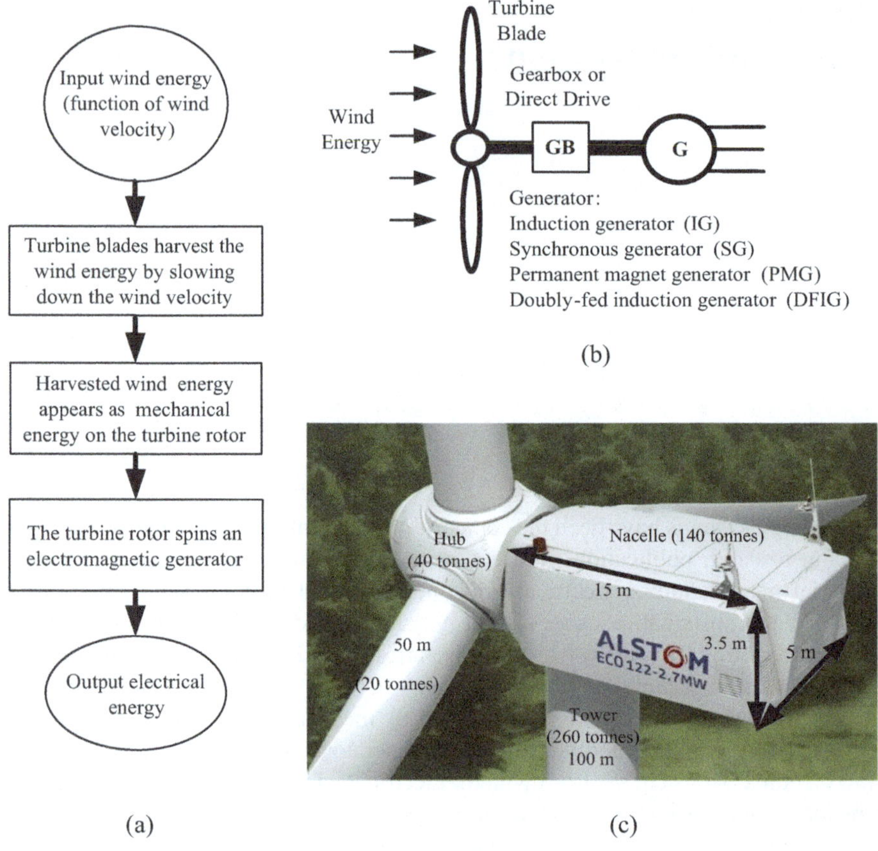

Fig. 1.1 Wind energy conversion. (**a**) Conversion from wind to electrical energy (**b**) Basic wind turbine power train components (**c**) Example 2.7 MW wind turbine [3]

The pitch control system changes the direction that the turbine blades are facing the wind. Therefore, by adjusting the pitch angle of a turbine, desired amount of energy is harvested from the wind. At zero pitch angle, the blades are fully facing the wind; hence, maximum wind energy is captured, and as the pitch angle increases, the energy harvested is reduced. The pitch angle is one of the tools that is used to control the power output of a wind turbine. The yaw control system rotates the turbine top parts (nacelle, blades, and hub) with respect to the turbine tower. In case the wind direction changes, the yaw system moves the turbine toward the wind velocity direction or out of it if desired. The generator low-output voltage is stepped up using a transformer to a level suitable for transmission. The turbine transformer is placed at the bottom of the turbine tower in some wind turbines. For the same power, the direct drive generator has usually a bigger diameter than a gearbox-coupled

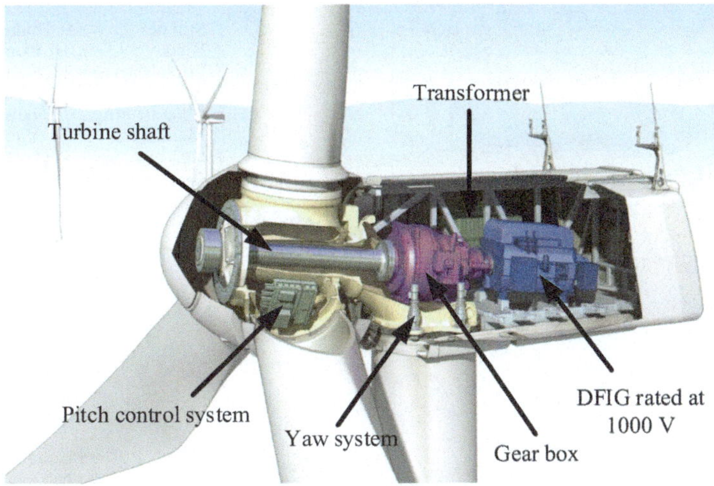

Fig. 1.2 Main components in an Alstom wind turbine [3]

generator, while its axial length is smaller. Figure 1.3 compares a 3.6 MW commercial wind turbine with gearbox-coupled generator manufactured by Siemens, SWT-3.6-107 3.6 MW, with a 2 MW wind turbine with direct generator manufactured by Enercon.

To collect the wind energy in a desired large amount, a number of wind turbines are normally clustered in a group and installed at an offshore or onshore "wind farm" or "wind park." An example of each is shown in Fig. 1.4.

There have been concerns about the onshore wind parks mainly due to turbine noise, visual impacts on the environment in which the turbine is installed, harms to birds, and also affecting neighboring property prices and limited land [7]. Therefore, much of the work in wind generation systems has shifted toward the offshore wind installations. In the offshore locations, the wind is usually stronger and steadier; hence, more power can be captured. However, it is worth noticing that the cost of installations in the offshore sites is normally higher than that of onshore farms.

1.2 Developments in the Wind Generation Systems

Early wind mills date back to the year 644 AD, used by Persians (Iranians) in the today's Persian-Afghan border region of Sistan and is shown in Fig. 1.5 [4].

The early wind mills were used locally for different purposes such as grinding wheat and pumping water but producing electricity. Around 1890s, first wind turbines were manufactured to produce electricity using DC generators. Design of wind turbines developed through the years by improved conversion power train such as mechanics, new generators, power electronics, and control systems. One of the

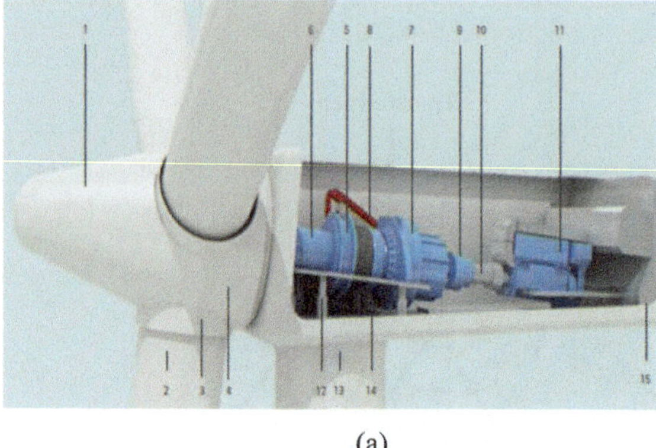

(a)

(b)

Fig. 1.3 Wind turbine components (**a**) Siemens SWT-3.6-107 3.6 MW wind turbine [5]. (**b**) Enercon E-82 2 MW wind turbine [6]

early modern wind turbine technology used induction generator directly connected to the grid known as direct online (DOL) as shown in Fig. 1.6. The DOL system was limited in energy conversion and required machine speed control via variable gearbox and blade pitch control. In the DOL, IG magnetizing current and reactive current is supplied by local grid. In an improved DOL connection, the reactive power was countered by variable (switched) capacitance connected in parallel to machine terminals.

The DOL systems were later improved using DFIG and power electronic converters shown in Fig. 1.7. The two converters in the scheme presented in Fig. 1.7 are so-called connected back to back. Each converter typically processes one-third of the rated DFIG power. The two converters decouple the rotor from machine output and allow the variable speed operation of DFIG within limited speed range. By injecting

(a)   (b)

Fig. 1.4 Wind farms. (**a**) Brazos onshore wind farm, USA [1]. (**b**) Walney offshore wind farm, UK [2]

Fig. 1.5 Early wind turbines were used by Persians in the Sistan region. (Photo courtesy of Google Map)

a variable frequency and amplitude voltage into the DFIG rotor, the machine output voltage and frequency are controlled to fixed values.

The wind turbine power trains based on DFIG are used in the today's systems. However, wind turbine systems with fully rated power electronic converters, shown in Fig. 1.8, are current industry practice for many wind generation systems. The two converters are voltage source converter (VSC) being connected back to back. For the current systems, the generator is usually an IG, a PM, or an SG. Due to varying

Fig. 1.6 DOL wind turbines

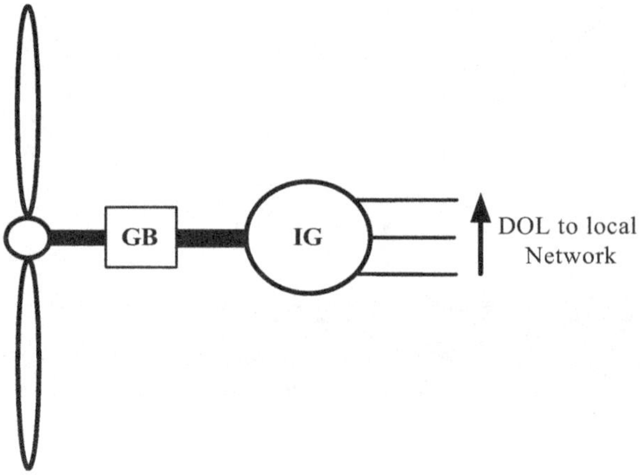

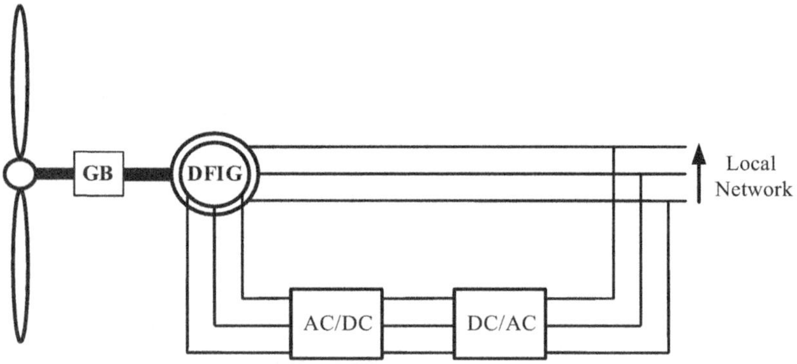

Fig. 1.7 DFIG and power electronic converters

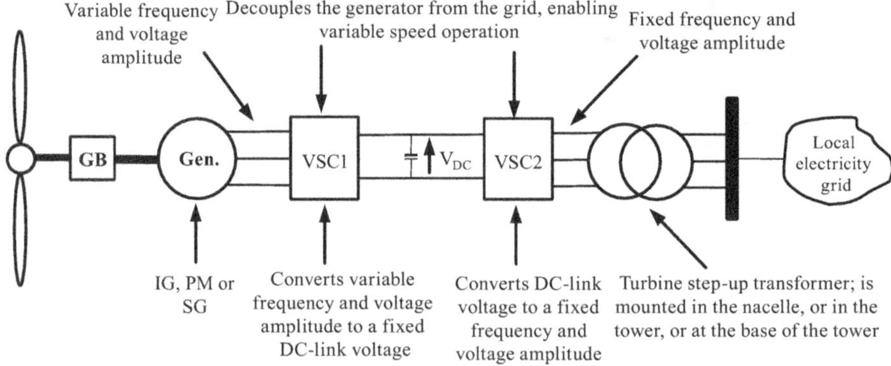

Fig. 1.8 Wind turbine power train with fully rated power electronic VSCs

nature of the wind velocity, the generator output voltage amplitude and frequency are variable. The generator side VSC, i.e., VSC1 in Fig. 1.8, converts generator variable frequency and voltage amplitude to a fixed DC-link voltage. VSC2 converts DC-link voltage to a fixed frequency and voltage amplitude at the transformer input. The VSCs decouple the generator from the grid, enabling variable speed operation. The DC-link voltage is maintained at its nominal value via grid-connected VSC2. The transformer steps up the voltage to a level suitable for transmission and sent it to the local grid.

Current studies have investigated many areas related to wind generation including, but not limited to, the control and stability of wind farms [8, 9], power electronic converters for the turbine generator power trains [10–12], and DC grids for wind generation [13–21]. Liu et al. [8] address wind farm stability and control using an impedance-based criterion where the VSC-based HVDC transmission is used in an offshore wind farm. For existing commercial wind generation systems, the aggregated power at the offshore substation is transmitted to the shore via either high-voltage AC (HVAC) or HVDC systems. Above a so-called break-even distance, HVDC will have lower costs than HVAC systems, providing the same transmission power capacity is considered. However, with existing technologies, the HVDC terminal station costs are generally more than those of AC systems, although this trend may change in the future with the development of higher voltage and current handling power electronic devices and converters. The choice between HVDC and AC system depends on several factors including voltage and power level, cost of devices, installations and labor costs, loss analysis, and control capabilities. However, trends suggest that as more systems are commercialized, HVDC converter station costs will continue to reduce, making it the favored choice in the future.

Feldman et al. [11] present a hybrid modular multilevel VSC for the HVDC with voltage control option for real and reactive power. A review of different converter topologies, control method, and efficiency is provided in [12] that include (i) a PMG connected to passive rectifier back-to-back with thyristor (SCR) converter, (ii) a PMG connected to a passive rectifier back-to-back with a voltage source inverter (VSI), (iii) a PMG connected to a passive rectifier back-to-back with a VSI and chopper in between, (iv) a PMG connected to two back-to-back VSC, (v) a DFIG with a rectifier and an SCR, (vi) a DFIG with two back-to-back VSC, (vii) a DFIG with a matrix converter, (viii) an IG connected to a passive rectifier and a VSC, and (ix) an SG connected to passive rectifier with a DC/DC boost converter and an inverter. Advantages and disadvantages of different generator-converter topology is also addressed in [12]. The HVDC system proposed in [14] reduces the number of VSC in the system by using a higher-rated VSC for a group of wind turbines. However, such a scheme results in lower (electrical) control functionality at the individual turbines. Veilleux [15] proposed a series of HVDC interconnection system using PMG, passive rectifier, and DC/DC converter for each turbine such that the offshore platform is eliminated. However, not having a platform offshore increases the system maintenance and serviceability. Moreover, the reliability of such system is at risk if the series connection fails. An HVDC system is proposed by ABB known as Windformer [16] that uses PMG connected to a passive rectifier which makes the Windformer unable of controlling the generator output voltage due to fixed PM excitation and lack of active power electronic converter; hence, it is infeasible to implement. The ABB Windformer was initially proposed in 2000 but has not been installed or investigated further.

Muyeen et al. [17] investigate dynamic and transient analyses of an offshore wind farm with on three-level neutral point clamped VSC HVDC transmission where the maximum power point tracking (MPPT) control of the wind turbine is implemented via VSCs, while [18–21] provide a review of HVDC systems based on different topologies, application, and models. Meyer et al. [22] propose a system of parallel DC interconnections where different topologies based on location of DC/DC converter are investigated and the topology where the DC/DC converters are placed on an offshore substation is chosen. The DC grid proposed by [22] employs DC/DC converters with a so-called single active bridge (SAB) topology [23]. Other topologies for high-voltage, high-power DC/DC converters have also been proposed based on dual active bridge (DAB) and resonant circuits [24–27].

DC/DC converter based on two resonant circuits is discussed by in [25], while Meyer [27] studies a series resonant DC/DC converter based on a modified topology of a three-phase series resonant circuit (SRC) system. Max [28] discusses design and control considerations for an SAB DC/DC converter where the losses are evaluated and a variable frequency control method results in low losses. Two topologies for resonant switched-capacitor DC/DC converter are proposed in [29], while [30] provides design of a bidirectional resonant 30 kW DC/DC converter where it is capable of DC fault isolation. A series and a cascade topology for multiple modules DC/DC converter is offered by [31] where the converter achieves high gains at high voltages. Ortiz et al. [32] present design of a high-power DC/DC converter where DAB and SRC configurations are investigated with zero current switching (ZCS) and the converter efficiency reaches 98.6%.

Different generator topologies and their comparison have also been proposed and demonstrated, although in small laboratory prototype scale [33, 34]. Chau et al. [33] proposed a hybrid excitation PM generator for the wind turbine application where a DC wound field winding is used in the stator in series with PM. The field winding regulates the machine fixed output voltage due to the PM excitation. However, the topology is complex and offers a low-voltage, low-power machine design. The hybrid excitation PM machine proposed in [34] has two stators and an outer rotor with 24 salient poles with no windings. The outer stator has a three-phase winding, while the inner stator uses both PM and DC wound field windings. The configuration is complex, and the machine is a low-voltage, low-power generator. Pillai et al. [35] discuss different generator topology for the wind turbine applications including cage and wound rotor IG, DFIG, and PM and investigate the SG in more details. Benefits presented for the SG in [35] include better power capability, transformer free since the HV SG is commercialized and widely in use, harmonic-free, and better fault current capability. However, on the disadvantages, it discusses increased drive train mass and complexity for the SG compared to other generator topologies. Margaris [36] studies a direct drive wound field SG connected to a grid via a passive rectifier, DC/DC converter, and a VSC where three different control strategies are discussed:

(i) constant output voltage by SG excitation, maximum power point tracking by DC/DC converter, and constant DC-link voltage using the VSC; (ii) power control using SG excitation, voltage control by DC/DC converter, and DC-link voltage control by VSC; and (iii) output voltage control by SG excitation, DC-link voltage control by DC/C converter, and power control by the VSC. Achilles et al. [37] discuss a model used for both dynamic and steady-state analysis of an SG connected to a grid using a passive rectifier, DC/DC converter to, and a VSC. Behnke [38] developed a reduced order model for the SG connected to a grid via VSCs to study system response to disturbances. Performance of wind turbine during grid voltage disturbance and speed variations is studied. Spooner [39] proposed direct drive light mass PM generator where the stator has no iron, while Brisset [40] discusses different topologies for axial-flux direct drive PM generator for wind turbine application. The machine stator has nine-phase concentric winding, and choice of configurations is based on finite element analysis (FEA). A comparison between large-scale geared and direct-drive generators for wind turbine applications has been presented in [41–43]; however, the comparison does not include a total wind generation scheme. Hence, the comparison is not comprehensive and cannot be generalized. Liserre [44] provides an overview of wind turbines in terms of generator technology, power electronic interface, and control strategies for the system. It also briefly discusses system issues including harmonics and faults. Most turbine generator systems are rated to an upper voltage limit of 1000 VAC, although there are some experimented or prototype installations using higher-voltage generators, such as an SG example operating at 13.8 kV [45].

Chapter 2
Wind Turbine Systems

2.1 Introduction

This chapter will present the equations used to analyze wind turbine systems and apply them to confirm the general operating characteristics of two commercial systems used in the wind industry, the Enercon E-82 2 MW turbine [6] and the Siemens SWT-3.6-107 3.6 MW turbine [5]. With this validation, the models are then used to identify the general industry operating principles of wind turbines, referred to herein as (i) the limited speed operating regime, and to propose an alternative (ii) high-speed operating regime.

2.2 Wind Turbine Power and the Betz Limit

A wind turbine is an energy converter that transforms the kinetic energy of the wind into mechanical energy that is then used as the prime mover to an electric generator. The electric generator subsequently converts mechanical to electrical energy. However, all the available energy in the wind cannot be harvested as observed by the German scientist Albert Betz [46] who showed that an ideal loss-less wind turbine can only convert 59.3% of the total wind energy. Referring to the schematic diagram of Fig. 2.1, the kinetic energy in a moving parcel of air is given by:

$$U = \frac{1}{2}\, m\, v^2 \quad \text{(Joules)} \tag{2.1}$$

which, considering the mass equation:

© Springer Nature Switzerland AG 2020
O. Beik, A. S. Al-Adsani, *DC Wind Generation Systems*,
https://doi.org/10.1007/978-3-030-39346-5_2

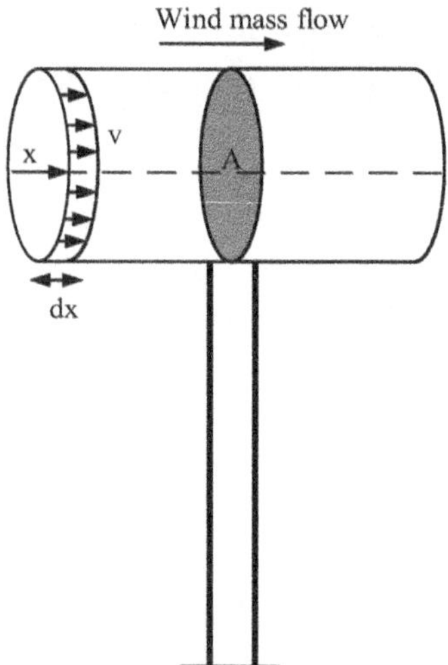

Fig. 2.1 Airflow with velocity of v through an area of A

$$m = \rho\, A\, x \ (\text{kg}) \tag{2.2}$$

becomes:

$$U = \frac{1}{2}\, m\, v^2 = \frac{1}{2}\, (\rho\, A\, x)\, v^2 \ (\text{Joules}) \tag{2.3}$$

where

m Parcel mass (kg)
v Parcel speed (m/s)
A Cross-sectional area (m^2)
ρ Air density (kg/m^3)
x The thickness of the parcel (m)

The power in the wind is obtained by taking the time derivative of the kinetic energy as follows:

$$P_w = \frac{dU}{dt} = \frac{1}{2}\, v^2\, \frac{dm}{dt} \ (\text{W}) \tag{2.4}$$

where the mass flow is:

$$\dot{m} = \frac{dm}{dt} = \rho\, A\, \frac{dx}{dt} \ (\text{kg/s}) \tag{2.5}$$

Linear velocity is:

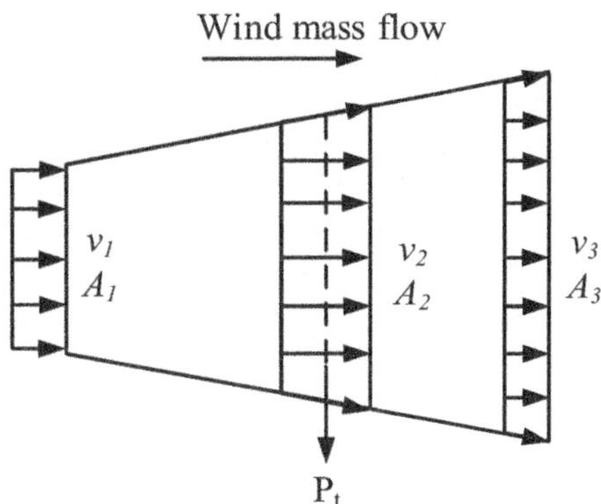

Fig. 2.2 Increase in cross-sectional area

$$\frac{dx}{dt} = v \ (\text{m/s})$$

Hence,

$$P_w = \frac{dU}{dt} = \frac{1}{2}\, \rho\, A\, v^3 \ (\text{W}) \tag{2.6}$$

The wind turbine harvests energy from the wind by slowing down the wind velocity. Therefore, with an unchanged mass flow, the cross-sectional area behind the turbine increases as a result of velocity reduction as shown in Fig. 2.2.

The power extracted from the wind is then related to the power difference of air before and after the turbine. Thus, referring to Fig. 2.2:

$$P_t = \frac{1}{2}\, \rho\, A_1\, v_1^3 - \frac{1}{2}\, \rho\, A_3\, v_3^3 = \frac{1}{2}\, \rho \left(A_1\, v_1^3 - A_3\, v_3^3 \right) (\text{W}) \tag{2.7}$$

For an unchanged mass flow:

$$\dot{m} = \rho\, A_1\, v_1 = \rho\, A_3\, v_3 \ (\text{kg/s}) \tag{2.8}$$

Therefore:

$$P_t = \frac{1}{2}\, \rho\, A_1\, v_1 \left(v_1^2 - v_3^2 \right) (\text{W}) \tag{2.9}$$

or in terms of mass flow:

$$P_t = \frac{1}{2}\, \dot{m} \left(v_1^2 - v_3^2 \right) (\text{W}) \tag{2.10}$$

The force that the air exerts on wind turbine can be expressed as:

$$F = \dot{m} \left(v_1 - v_2 \right) (\text{N}) \tag{2.11}$$

This force is counteracted by an equal force exerted by the wind turbine on the airflow. Therefore:

$$P_t = F \, v_2 = \dot{m} \left(v_1 - v_2 \right) v_2 \; (\text{W}) \tag{2.12}$$

Equating Eqs. (2.10) and (2.12) results in:

$$v_2 = \frac{\left(v_1 + v_3 \right)}{2} \; (\text{m/s}) \tag{2.13}$$

Therefore:

$$\dot{m} = \rho \, A_2 \, v_2 = \rho \, A_2 \left(\frac{v_1 + v_3}{2} \right) \; (\text{kg/s}) \tag{2.14}$$

Thus, the wind turbine power can be expressed as:

$$P_t = \frac{1}{2} \, \dot{m} \left(v_1^2 - v_3^2 \right) = \frac{1}{4} \, \rho \, A_2 \left(v_1^2 - v_3^2 \right) \left(v_1 + v_3 \right) \; (\text{W}) \tag{2.15}$$

According to Eq. (2.6), the power in the wind can be expressed as:

$$P_w = \frac{1}{2} \, \rho \, A_1 \, v_1^3 \; (\text{W}) \tag{2.16}$$

The ratio of the wind turbine power to the available power in the wind is called the wind turbine power coefficient C_P:

$$C_P = \frac{P_t}{P_w} \tag{2.17}$$

Using Eqs. (2.15) and (2.16), the wind turbine power coefficient is calculated from:

$$C_P = \frac{1}{2} \left(1 + \frac{v_3}{v_1} \right) \left[1 - \left(\frac{v_3}{v_1} \right)^2 \right] \tag{2.18}$$

The turbine power coefficient variation with respect to the ratio of velocities before and after the turbine is plotted in Fig. 2.3. Here, C_P is maximum at a velocity ratio that is $v_3/v_1 = 1/3$; hence, the maximum turbine power coefficient is:

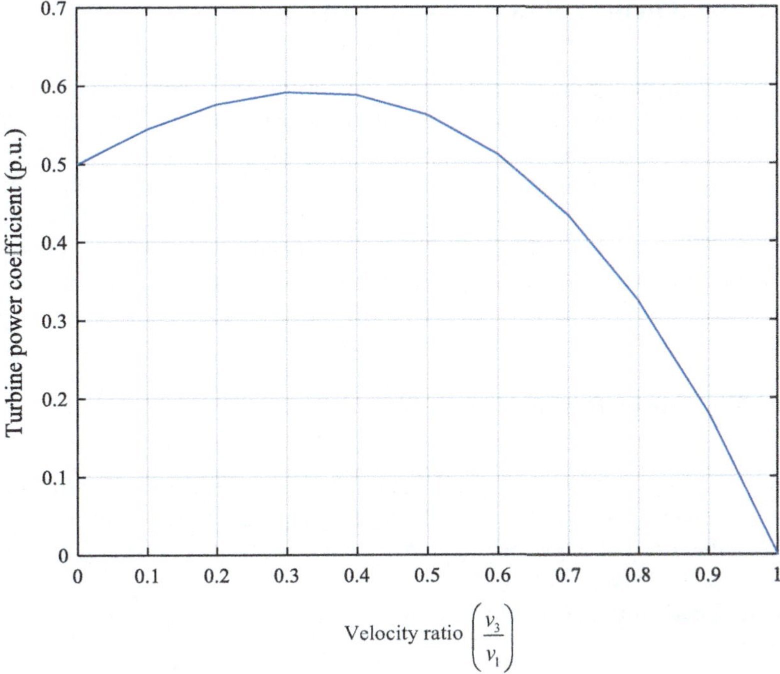

Fig. 2.3 Turbine power coefficient variation with respect to the ratio of wind velocities before and after the turbine

$$C_P^{\mathrm{max}} = \frac{16}{27} = 0.593 \text{ per unit} \qquad (2.19)$$

This is the "Betz limit" or "Betz factor" as referred to previously. Therefore, an ideal loss-less wind turbine can harvest a maximum of 59.3% of the wind energy.

The net extractable turbine power, therefore, is expressed as:

$$P_t = \frac{1}{2} \rho A v^3 C_P \text{ (W)} \qquad (2.20)$$

For nonideal turbines, where there is further loss of energy conversion, the turbine power coefficient is modified.

Using the turbine rotational speed ω (rad/s), the turbine torque, T_t, is obtained from:

$$T_t = \frac{P_t}{\omega} = \frac{1}{2} \rho A v^3 \frac{C_P}{\omega} = \frac{1}{2} \rho A v^3 C_\tau \text{(N.m)} \qquad (2.21)$$

where $C_\tau = \frac{C_P}{\omega}$ is the turbine torque coefficient.

2.3 Wind Turbine Power Coefficient

The wind turbine power coefficient defines the portion of the power that is extracted from the wind. The power coefficient, therefore, depends on the aerodynamic design of the wind turbine. In order to obtain the turbine power coefficient, different measurements at various wind velocities are taken and the turbine power coefficient estimated from the post process analysis. Analytical methods may also be applied to model the turbine power coefficient. Normally, manufacturers provide the turbine power along with the power coefficients at different velocities.

2.3.1 Tip Speed Ratio

The turbine power coefficient, C_P, is normally expressed as a function of two parameters: blade pitch angle, β, and blade tip speed ratio, λ. The blade pitch angle is defined as the angle between the blade cross-section chord and the plane of rotation as depicted in Fig. 2.4.

The tip speed ratio is defined as follows:

$$\lambda = \frac{u}{v} = \frac{R\,\omega}{v} \tag{2.22}$$

where

u is the tangential velocity of the blade tip.
R is the rotor radius ($\approx$ blade length).

The rotor radius includes the hub radius plus the blade length, but in large wind turbines, the blade length is much longer than the hub radius; hence, it is acceptable to assume that R is the total blade length.

The turbine power coefficient is different for various turbine designs and configurations. Typical wind turbine power coefficient (C_P) and also torque coefficient (C_τ) curves for different turbine configurations are presented in Fig. 2.5 [4]. The upper limit for the turbine power coefficient is the Betz limit. As it can be seen, the three-

Fig. 2.4 Blade pitch angle

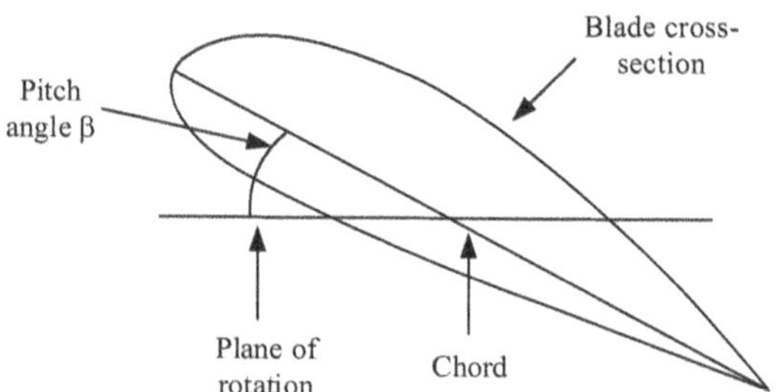

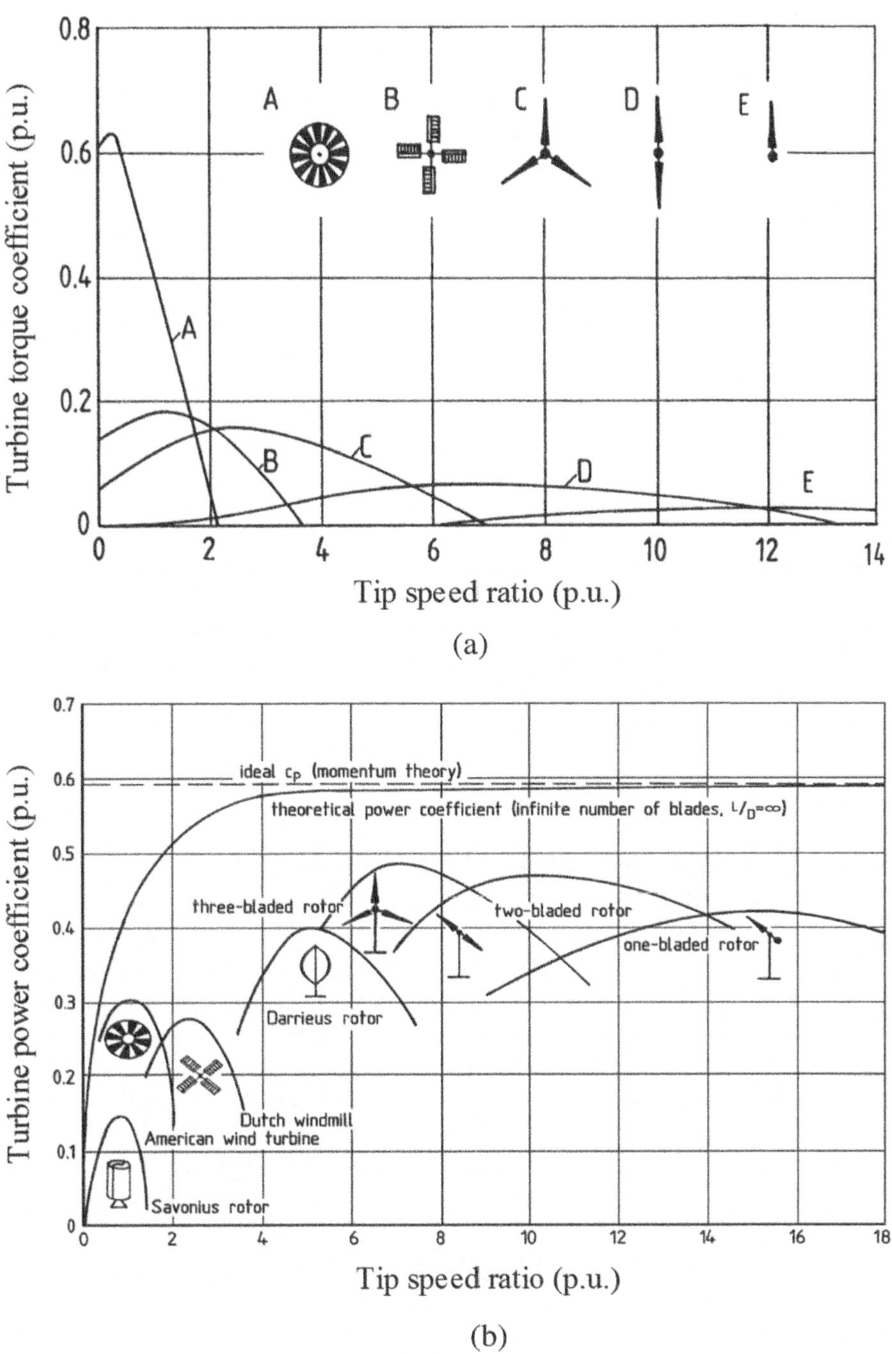

(a)

(b)

Fig. 2.5 Wind turbine power and torque coefficients' curves [4] (**a**) Torque coefficient. (To be replaced with a drawing) (**b**) Power coefficient

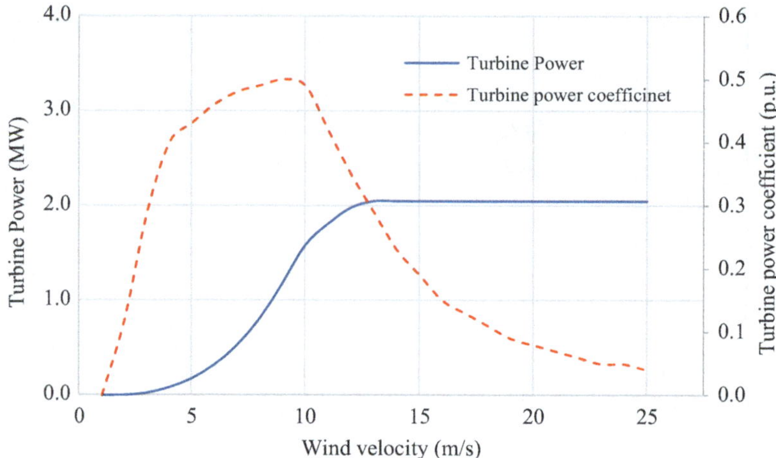

Fig. 2.6 Turbine power and power coefficient for Enercon E-82

blade wind turbine has the highest power coefficient among various designs; hence, modern wind turbines generally use three-blade designs.

One difficulty in assessing manufacturer's turbines is that the relevant turbine pitch blade angles at which the power coefficients are obtained are not usually provided. As an example, the turbine power and the power coefficient for Enercon E-82 2 MW are presented in Fig. 2.6.

2.3.2 *Mathematical Expression for Wind Turbine Power Coefficient*

The power coefficient of wind turbines can be approximated by nonlinear functions as discussed in great detail by O. Wasynczuk in 1981 [49] and P. M. Anderson [50] in 1983, who reported the power coefficient for an MOD-2 wind turbine that is an upwind, two-blade, 2500 kW wind turbine installed and tested in the USA [51]. Upwind is a term used when the wind turbine is facing the wind in front of the electric machine, while for downwind turbines, the electric machine is behind the turbine in terms of wind flow. The equation for the power coefficient from [49, 50] is:

$$C_P = 0.5 \left(\Gamma - 0.022\,\beta^2 - 5.6 \right) e^{-0.17\Gamma} \qquad (2.23)$$

where

$$\Gamma = \frac{v}{\omega}$$

S. Heier [47] proposed the following model for the wind turbine power coefficient:

$$C_P\left(\lambda,\beta\right) = c_1\left(c_2 - c_3\,\beta - c_4\,\beta^x - c_5\right)\mathrm{e}^{-c_6} \tag{2.24}$$

Using parameters stated in Eqs. (2.25), (2.24) fits to Eq. (2.23):

$$
\begin{aligned}
c_1 &= 0.5 \qquad c_2 = \frac{v}{\omega} \qquad c_3 = 0 \quad c_4 = 0.022 \\
c_5 &= 5.6 \qquad c_6 = 0.17\,\frac{v}{\omega} \qquad\quad x = 2
\end{aligned}
\tag{2.25}
$$

The power coefficient following parameters has been proposed by S. Heier [47], Z. Lubosny [48], and B. Almang [52]:

$$
c_1 = 0.5 \qquad c_2 = \frac{116}{\lambda_i} \qquad c_3 = 0.4 \qquad c_4 = 0 \qquad c_5 = 5 \qquad c_6 = \frac{21}{\lambda_i}
$$

$$
\frac{1}{\lambda_i} = \frac{1}{\lambda + 0.08\,\beta} - \frac{0.035}{\beta^3 + 1}
\tag{2.26}
$$

However, a modified version of the turbine power coefficient and parameters is derived here as follows:

$$C_P\left(\lambda,\beta\right) = c_1\left(c_2 - c_3\,\beta - c_4\,\beta^x - c_5\right)\mathrm{e}^{-c_6} + c_7\,\lambda$$

$$c_1 = 0.5176\,,\quad c_2 = \frac{116}{\lambda_i}\,,\quad c_3 = 0.4\,,\quad c_4 = 0\,,\quad c_5 = 5\,,\quad c_6 = \frac{21}{\lambda_i}\,,\quad c_7 = 0.0068$$

$$\frac{1}{\lambda_i} = \frac{1}{\lambda + 0.08\,\beta} - \frac{0.035}{\beta^3 + 1}$$

$$\tag{2.27}$$

The maximum of the turbine power coefficient stated in Eq. (2.24) when applying parameters in Eq. (2.26) is 0.41 p.u., while using Eq. (2.27), a maximum of 0.48 p.u. is achieved. Note that the turbine power coefficient cannot be higher than the Betz limit, i.e., 0.593 p.u. Today's wind turbine design's trend suggests C_P values in the range of 0.47–0.5 p.u. Therefore, Eq. (2.27) for the wind turbine power coefficient is used for further analysis.

The power coefficients can be different for various turbines. The C_P equations provide an approximation to the actual turbine power coefficient; hence, the equations and their parameters can vary depending on the design of the turbine. As an example, the maximum value of C_P for Enercon E-82 2 MW wind turbine is 0.5 that

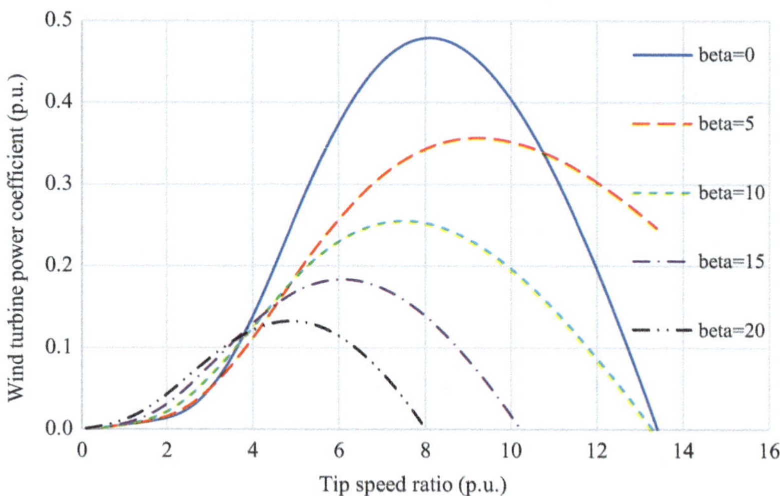

Fig. 2.7 Turbine power coefficient curves

does not match the maximum C_P in Eq. (2.27). Other methods of approximating the turbine power coefficient such as curve fitting [53] and exploiting the drag and lift characteristics of the turbine [48] have been proposed. However, if the C_P data and pitch angles for different wind velocities and rotor speeds are provided by manufacturers, analysis of the wind turbine should rely on real data. But in many cases, the manufacturers only provide a curve of power coefficient versus wind velocity without pitch angles and rotor speeds; thus, approximation equations are used.

By varying the tip speed ratio, λ, the turbine power coefficient curves for different pitch angles using Eq. (2.27) are plotted in Fig. 2.7. The turbine power coefficient curve at $\beta = 0$ achieves the highest value compared to curves at other pitch angles; hence, a wind turbine harvests the highest power when it operates at zero pitch angle.

2.4 Wind Turbine Operation

By replacing wind velocity from Eq. (2.22) to Eqs. (2.20) and (2.21), the turbine power and torque are, respectively, expressed as:

$$P_t = \frac{1}{2} \rho \pi R^5 \omega^3 \frac{C_P(\lambda, \beta)}{\lambda^3} \ (\text{W}) \tag{2.28}$$

$$T_t = \frac{P_t}{\omega} = \frac{1}{2} \rho \pi R^5 \omega^2 \frac{C_P(\lambda, \beta)}{\lambda^3} \ (\text{N.m}) \tag{2.29}$$

where the air density in normal conditions is $\rho = 1.225$ (kg/m^3). The turbine power and torque are cubic and square function of the turbine rotor speed (and wind velocity), respectively.

Using blade a length of $R = 41$ (m) for the Enercon E-82 2 MW, the wind turbine power and torque characteristics versus rotor speed for $\beta = 0$ and at different wind velocities are plotted in Fig. 2.8. The so-called maximum power point tracking (MPPT) of the turbine is achieved by controlling the turbine system such that it operates at the maximum points of the power curves shown in Fig. 2.8a, i.e., every point on the MPPT corresponds to a maximum power being extracted from the wind at a specific wind velocity. Commercial wind turbines are normally rated to operate over a variable wind velocity range that defines the minimum and maximum operating points for the turbine. The minimum wind velocity at which the turbine starts delivering power is called the "cut-in velocity," and the maximum wind velocity for turbine operation is called the "cut-out velocity." Although the output power of the wind turbine varies with wind velocity, industrial turbines are generally designed and controlled to deliver a constant power referred to as "rated power" over a variable range of wind velocity. Rated here usually refers to the system mechanical, thermal, and electromagnetic capabilities.

For the purpose of comparison and further discussion, the wind velocity at which the turbine starts delivering rated power will be defined as the "rated power base velocity," or simply the "rated velocity." This definition is industry practice and is clarified by reference to the power versus wind velocity characteristic for the Siemens SWT-3.6-107 3.6 MW wind turbine, as illustrated in Fig. 2.9, while Table 2.1 presents the Siemens SWT-3.6-107 3.6 MW wind turbine operational parameters.

Therefore, using Eq. (2.20) and according to Fig. 2.9, the power versus wind velocity characteristic of a wind turbine is mathematically expressed as follows:

$$P_t = \begin{cases} 0 & v \leq v_{\text{cut-in}} \\ \frac{1}{2}\,\rho\,\pi\,R^5\,\omega^3\,C_P(\lambda,\beta) & v_{\text{cut-in}} \leq v \leq v_{\text{rated}} \\ P_{\text{nominal}} & v_{\text{rated}} \leq v \leq v_{\text{cut-out}} \\ 0 & v \geq v_{\text{cut-out}} \end{cases} \quad (\text{W}) \qquad (2.30)$$

Figure 2.10 illustrates turbine power coefficient curves with respect to wind velocity at different rotor speeds for the Enercon E-82 wind turbine (rated at 2 MW). Using these curves, turbine power with respect to wind velocity variation at different rotor speeds is plotted in Fig. 2.11, from which it is seen that in order to extract the maximum power at a specific wind velocity, the rotor speed should be at a certain value which is referred to as the "optimum rotor speed." Therefore, for each wind velocity, there is an optimum rotor speed, and the wind turbine is controlled to operate at optimum rotor speeds.

Referring to Fig. 2.9, from cut-in to rated wind velocity, the turbine is controlled via MPPT; hence, power varies with wind velocity as defined by:

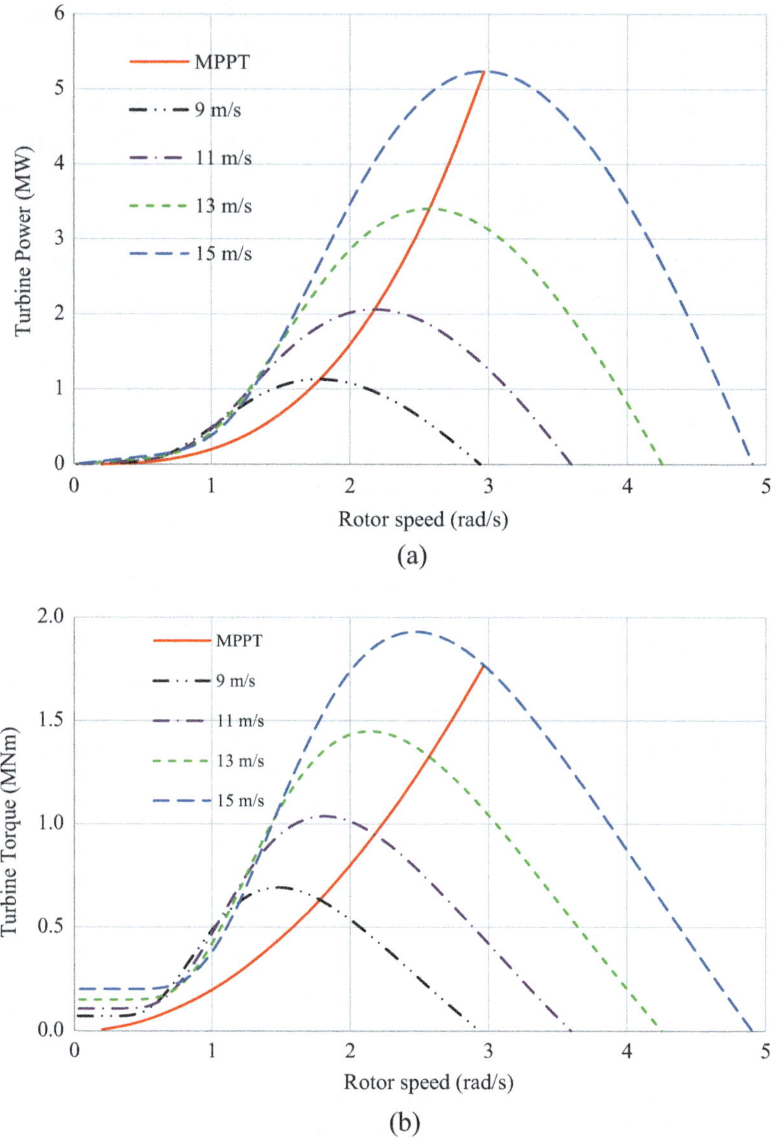

Fig. 2.8 Wind turbine power and torque characteristics for $\beta = 0$ (**a**) Turbine power curves (**b**) Turbine torque curves

$$P_t = \frac{1}{2}\, \rho\, \pi\, R^2\, v^3\, C_P^{\max}\ \ (\text{W}) \tag{2.31}$$

where $C_P^{\max}$ is the maximum turbine power coefficient.

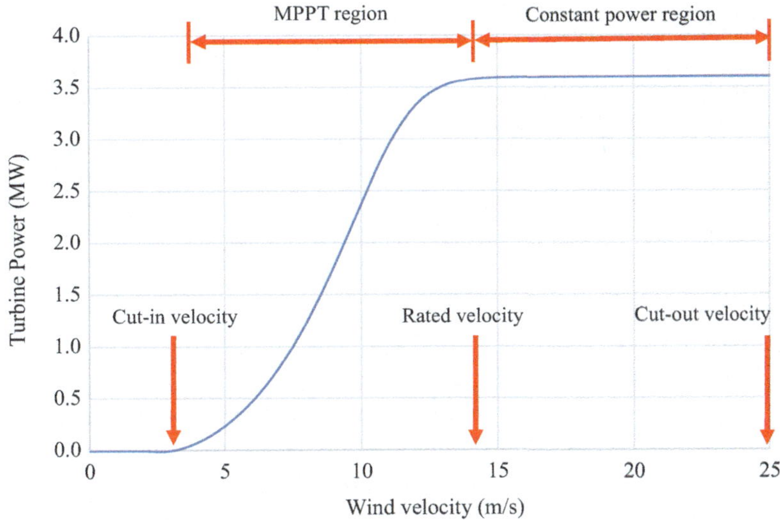

Fig. 2.9 Siemens SWT-3.6-107 3.6 MW wind turbine power curve with respect to wind velocity

Table 2.1 Operational parameters for the Siemens SWT-3.6-107 3.6 MW turbine

Parameter	Symbol	Value
Optimum tip speed ratio	λ_{opt}	8.1
Maximum turbine power coefficient	C_P^{max}	0.48
Wind velocity (m/s)	v	0–25
Cut-in velocity (m/s)	v_{cut-in}	3–5
Rated velocity (m/s)	v_{rated}	13–14
Cut-out velocity (m/s)	$v_{cut-out}$	25
Rated power (MW)	$P_{nominal}$	3.6
Air density (kg/m^3)	ρ	1.225

Hence, the turbine power is a cubic function of the wind velocity. For wind velocities above the rated velocity, the wind turbine is controlled to deliver a constant power, as shown in Fig. 2.9.

Commercial manufacturers use a control method that is explained as "limited speed operation" region. The "limited speed operation" is regarded as the standard industry practice to control turbines to deliver fixed power. However, a new control method for operating at wind velocities above the rated velocity is discussed in this book and referred to as "high speed operation." This control function allows the wind turbine to operate at higher rotor speeds while delivering a fixed power output. Note that for both limited and high speed operation regions, the wind turbine is controlled to ideally operate with MPPT from cut-in to the rated velocity. Hence, only when above the rated velocity will the control methods differ for limited and high speed operation.

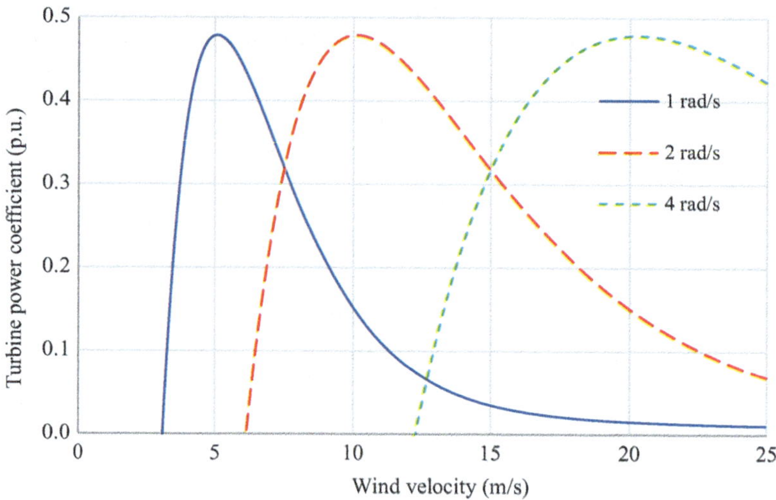

Fig. 2.10 Turbine power coefficient with respect to wind velocity variations at different rotor speeds for $\beta = 0$

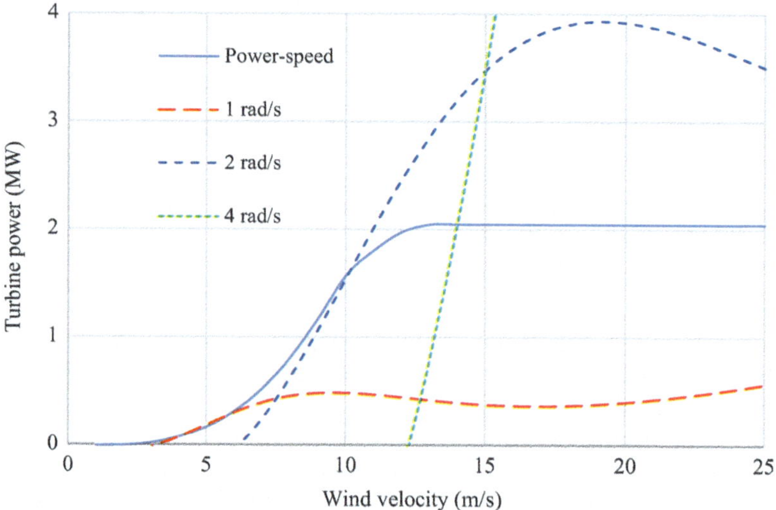

Fig. 2.11 Turbine power curves with respect to wind velocity variations at different rotor speeds (in rad/s) for $\beta = 0$

2.4.1 Limited Speed Operation

For the limited speed operation scheme, the turbine blade pitch angle is controlled such that turbine operates on the MPPT curve from cut-in up to rated velocity. Therefore, as the wind velocity increases, the turbine rotor speed and hence the

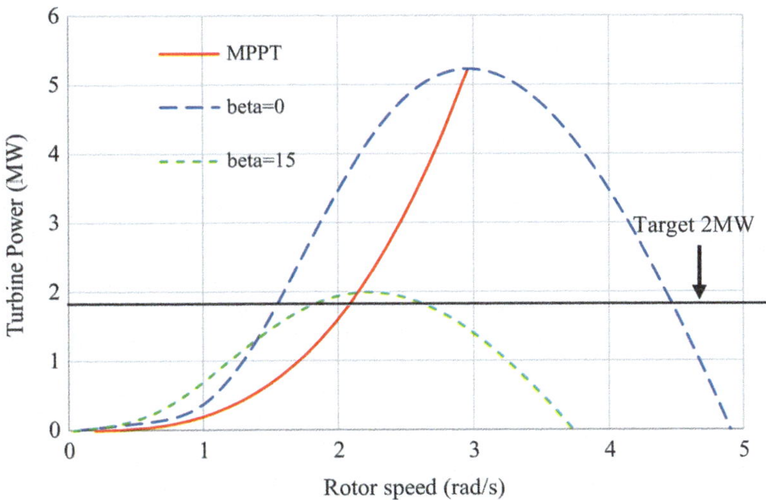

Fig. 2.12 Turbine power curves at wind velocity 15 m/s

power extracted from the wind increase. However, once at rated wind velocity and above, the turbine speed is maintained around a fixed upper speed, referred to here as the "turbine rated speed," or simply "rated speed." For wind velocities above rated velocity, the turbine blade pitch angle is controlled to maintain turbine speed at rated, and the turbine delivers a fixed rated power. It might be practically infeasible to control the turbine to operate exactly at the rated speed, and a limited speed variation of 10% around the turbine rotor rated speed is usually allowed.

In order to illustrate operation of a wind turbine in the limited speed operation region, the following case using the Enercon E-82 commercial wind turbine is analyzed and presented. The rated wind velocity for the turbine under study is 13 (m/s). Figures 2.12, 2.13, and 2.14 show turbine power curves at different blade pitch angles and for velocities above rated. If the blade pitch angle is maintained when the wind velocity increases to 15 (m/s), the turbine rotor speed, and hence output power, increases. However, to keep a constant output power and control the rotor at rated speed, the blade pitch angle is changed from zero to $\beta = 15°$, as shown in Fig. 2.12. For higher wind velocities, the blade pitch angle increases further to keep the turbine power and the speed around their nominal rated values. For example, at 19 and 25 m/s, the blade pitch angle is increased to $\beta = 25°$ and $\beta = 33°$, respectively, as shown in Figs. 2.13 and 2.14, respectively. Figure 2.15 illustrates the turbine per unit power-speed and torque-speed characteristics for the limited speed operation control regime. The speed variation around rated speed is about 10%, as explained previously arising from the result of practical implementation of the rotor blade pitch angle control.

It may be inferred from Fig. 2.15 that from cut-in to rated wind velocity, the turbine power is a cubic function of the turbine speed while the turbine torque is a square function of turbine speed, as per Eqs. (2.28) and (2.29), respectively. For

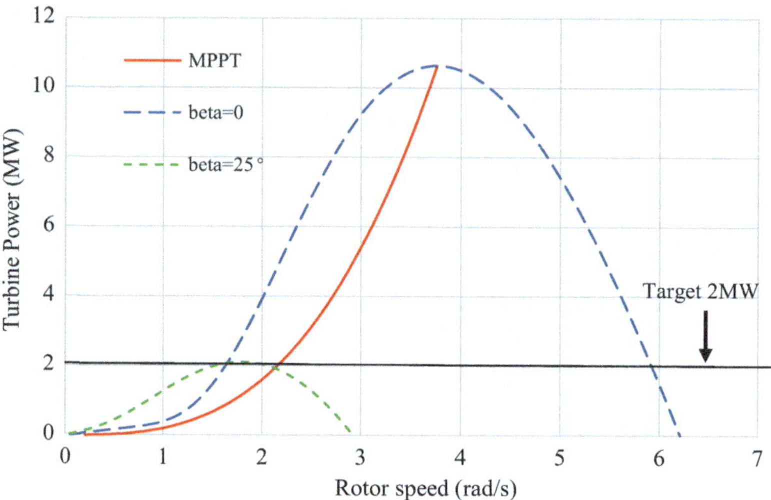

Fig. 2.13 Turbine power curves at 19 (m/s)

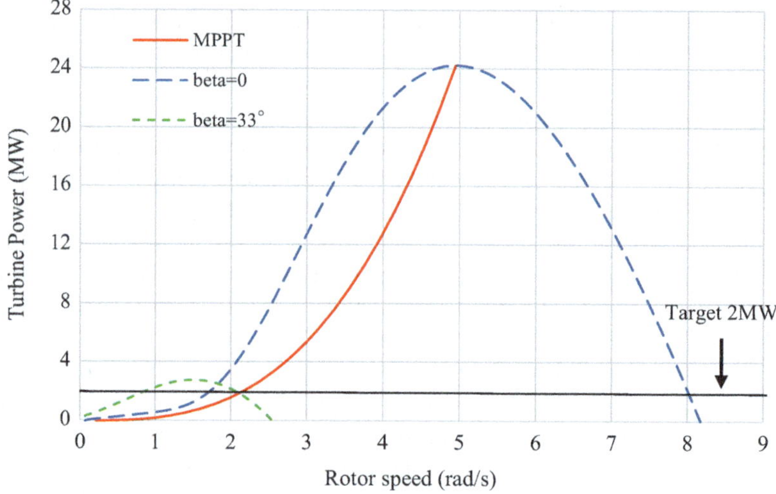

Fig. 2.14 Turbine power curves at 25 (m/s)

wind velocities higher than rated, power is controlled to be nearly constant while the turbine speed has small changes around the rated speed, which yield a small torque variation. Turbine rotor speed with respect to the wind velocity variations for limited speed operation is plotted in Fig. 2.16.

Using Eqs. (2.28 and 2.29) and according to Fig. 2.16, following expressions for power- and torque-speed characteristics of a turbine in the limited speed operation are derived:

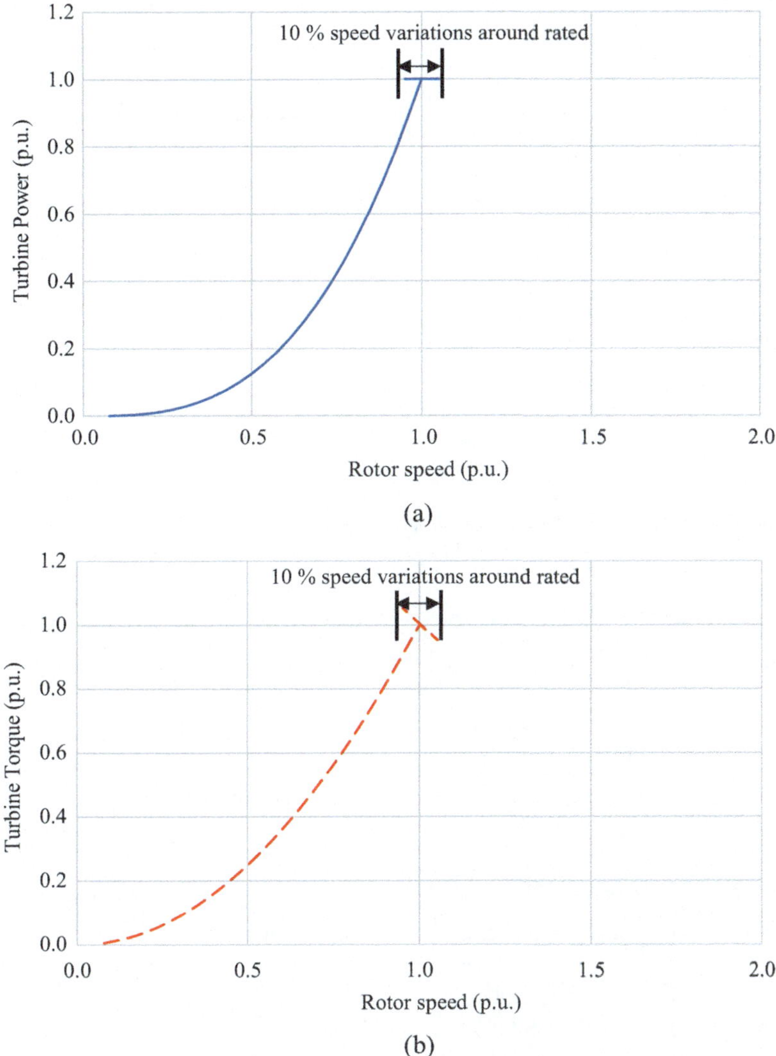

Fig. 2.15 Per unit power and torque characteristics of the wind turbine when operating in limited speed region. (**a**) Power speed (**b**) Torque speed

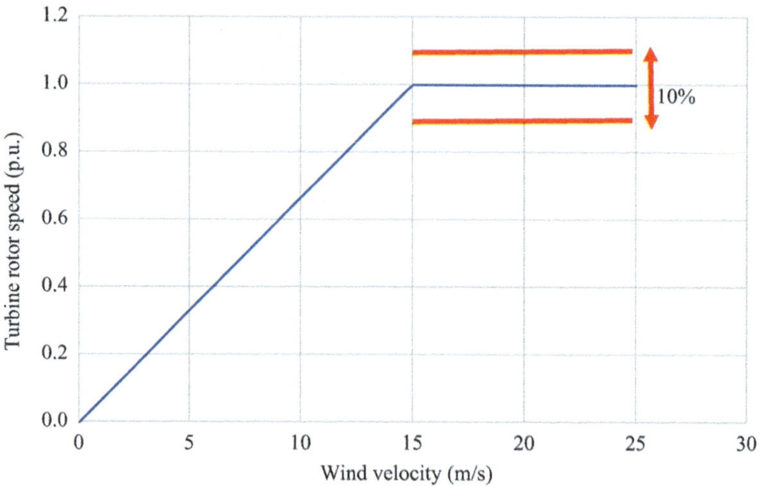

Fig. 2.16 Rotor speed for the limited speed operation region

$$
P_t = \begin{cases}
0 & \omega \le \omega_{\text{cut-in}} \\[2mm]
\dfrac{1}{2}\,\rho\,\pi\,R^5\,\omega^3\,\dfrac{C_P(\lambda,\beta)}{\lambda^3} & \omega_{\text{cut-in}} \le \omega \le \omega_b \\[2mm]
P_{\text{nominal}} & \omega = \omega_b\,(\pm\alpha\%) \\[2mm]
0 & \text{otherwise}
\end{cases} \quad \text{(W)} \qquad (2.32)
$$

$$
T_t = \begin{cases}
0 & \omega \le \omega_{\text{cut-in}} \\[2mm]
\dfrac{1}{2}\,\rho\,\pi\,R^5\,\omega^2\,\dfrac{C_P(\lambda,\beta)}{\lambda^3} & \omega_{\text{cut-in}} \le \omega \le \omega_b \\[2mm]
\dfrac{P_{\text{nominal}}}{\omega_b\,(\pm\alpha\%)} & \omega = \omega_b\,(\pm\alpha\%) \\[2mm]
0 & \text{otherwise}
\end{cases} \quad \text{(N.m)} \qquad (2.33)
$$

where

$\omega_{\text{cut-in}}$ is the rotor speed at cut-in wind velocity (rad/s).
ω_b is the rotor rated speed (rad/s).
α is the percentage of the speed variations around the rated speed.

2.4.2 High Speed Operation

Commercial wind turbines normally operate in the limited speed operation region. The high speed operating region is a new control concept, the operation of which is presented in this book. For wind velocities up to rated, the operation of the wind turbine in high speed region is as that of the limited speed operation scheme. The only difference in the two schemes is when the turbine operates above the rated wind velocity. Therefore, from cut-in to the rated wind velocity, the blade pitch angle is controlled such that maximum power is extracted from the wind and the turbine operates on the MPPT curve. However, for wind velocities above rated, the turbine speed is allowed to increase while harvesting rated power from the wind by controlling the blade pitch angle. By way of example, consider the results shown in Fig. 2.12. At a wind velocity of 15 (m/s), the blade pitch angle is increased to $15°$ to maintain the turbine power output at 2 MW, i.e., control in the limited speed region. However, if the blade pitch angle is held at zero, the turbine rotor would accelerate to $\omega = 4.5$ (rad/s) if the power output was maintained at 2 MW. This speed is obviously higher than that at rated wind velocity. The rotor speed at rated wind velocity for the Enercon E-82 is $\omega = 2$ (rad/s). Similarly, to harvest 2 MW from the turbine at velocities of 19 m/s (Fig. 2.13) and 25 m/s (Fig. 2.14) and $\beta = 0$, the turbine rotor speed would have to be controlled at $\omega = 5.9$ (rad/s) and $\omega = 8$ (rad/s), respectively. The per unit power-speed and torque-speed characteristics of the wind turbine when operating in the high operation speed region are plotted in Fig. 2.17, where it can be seen that above the rated speed, turbine torque reduces as a result of increased rotor speed at constant power. If, to be consistent with the limited speed operating scheme, the rated turbine rotor speed is referred to as 1 p.u., then for the Enercon E-82 studied here, maintaining the turbine blade pitch angle at zero results in the rotor speed increasing up to 2.66 p.u. at the wind cut-out velocity of 25 m/s. An alternative high speed operating regime would be to limit the increase in speed above the rated speed via control of the turbine blade pitch angle. This angle could be modified to limit speed, as illustrated in Fig. 2.17 showing the power-speed (a) and torque-speed (b) profiles for both of the high speed (HS) operating regimes, i.e., without (i) and with (ii) blade pitch control. In this example, the highest turbine rotor speed is 2.66 p.u. for HS(i) and 1.67 p.u. for HS(ii).

Figure 2.18 shows rotor speed with respect to wind velocity variation for both the limited and the two high speed operation regimes. According to Fig. 2.17, following expressions for power- and torque-speed characteristics of a turbine in the high speed operation regime is derived:

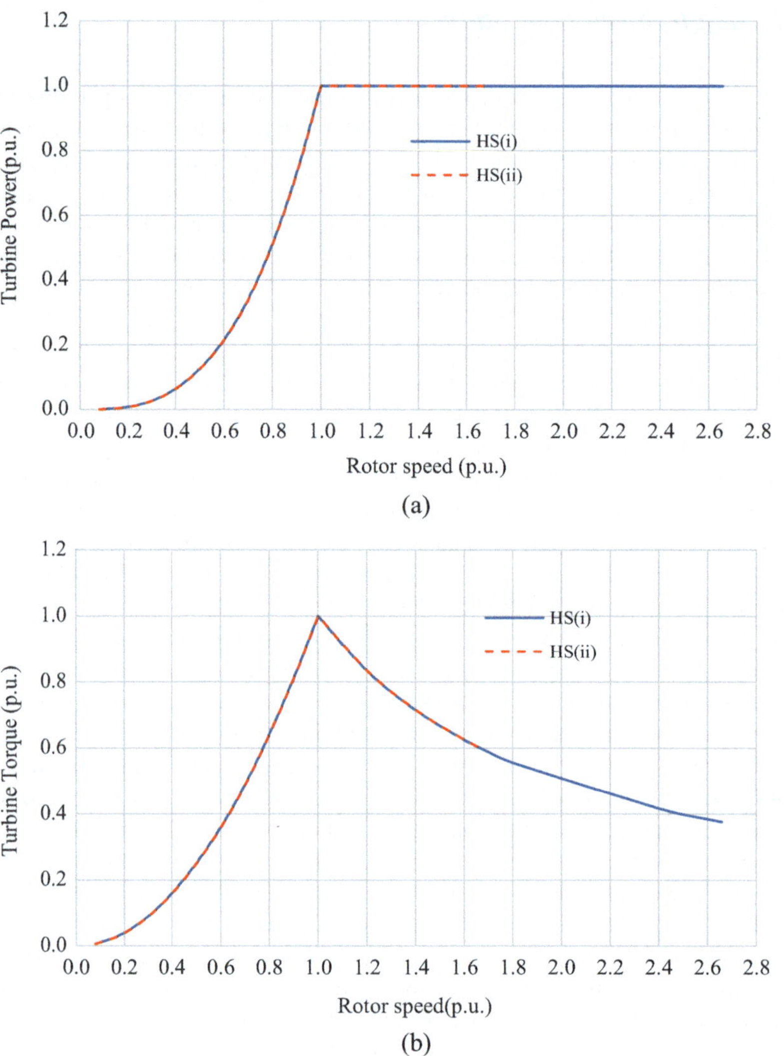

(a)

(b)

Fig. 2.17 Per unit power and torque characteristics of the wind turbine when operating in high speed region (**a**) Power versus speed (**b**) Torque versus speed

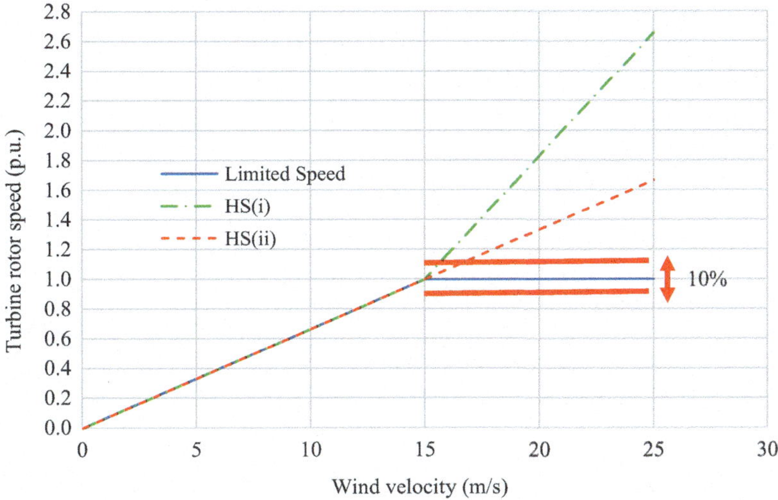

Fig. 2.18 Rotor speed for the limited and high speed operation regions

$$P_t = \begin{cases} 0 & \omega < \omega_{\text{cut-in}} \\ \dfrac{1}{2}\,\rho\,\pi\,R^5\,\omega^3\,\dfrac{C_P(\lambda,\beta)}{\lambda^3} & \omega_{\text{cut-in}} \le \omega < \omega_b \\ P_{\text{nominal}} & \omega_b \le \omega \le \omega_{\text{cut-out}} \\ 0 & \omega > \omega_{\text{cut-out}} \end{cases} \quad (\text{W}) \qquad (2.34)$$

$$T_t = \begin{cases} 0 & \omega < \omega_{\text{cut-in}} \\ \dfrac{1}{2}\,\rho\,\pi\,R^5\,\omega^2\,\dfrac{C_P(\lambda,\beta)}{\lambda^3} & \omega_{\text{cut-in}} \le \omega < \omega_b \\ \dfrac{P_{\text{nominal}}}{\omega_b\,(\pm\alpha\%)} & \omega_b \le \omega \le \omega_{\text{cut-out}} \\ 0 & \omega > \omega_{\text{cut-out}} \end{cases} \quad (\text{N.m}) \qquad (2.35)$$

The high speed region of operation is not usually implemented by high-power wind turbine system manufacturers; however, it may be interesting from the machine point of view due to the reduced torque at higher speeds and possible control simplifications. A major challenge for this operating regime is the impact of higher rotor speeds on the wind turbine mechanics, the assessment of which is out of the scope of this book. Although the mechanical implementation of the high speed operating regime is not further investigated, the impact of the scheme on generator power-speed and torque-speed specifications is discussed for completeness.

Chapter 3
DC Wind Generation System

3.1 Overview of the Walney Offshore Wind Farm

The Walney wind farm, shown in Fig. 3.1, is at present one of the largest offshore wind farms in the UK, located in the Irish Sea about 15 km off the UK coast near Barrow-in-Furness, Cumbria [2]. The total installation consists of two wind farms, Walney 1 and Walney 2, each with 51 turbines giving 102 turbines in total.

The wind turbine arrangements for Walney 1 and Walney 2 are shown in Fig. 3.2. The Walney wind farm turbines are connected to two offshore substations, one for each farm, via undersea AC cables. The offshore substations collect the output from all turbines at 33 kV and step up the voltage to 132 kV for transmission to the onshore substation using AC cables buried under the sea. The Walney 1 offshore substation is connected to the UK national grid at Heysham via a 44.4 km long buried cable.

The Walney wind farm is claimed to be operational for 25 years from the start of 2012. Table 3.1 presents summary details of the Walney wind farm, while Table 3.2 presents details of the Siemens 3.6 MW wind turbine SWT-3.6-107 which is used in Walney 1 wind farm. The Walney 1 wind farm is chosen as a representative commercial benchmark against which the new high-voltage scheme is assessed.

Figure 3.3 simply illustrates the electrical installation of the Walney 1 wind farm showing the major system components. The power train in this system uses an induction generator (IG) that is interfaced to the fixed frequency grid via two fully rated VSCs (Converter 1 and Converter 2) in a back-to-back configuration. The two VSCs decouple the IG output from the fixed frequency, (essentially) fixed voltage grid facilitating variable speed operation of the wind turbine. The maximum rated output voltage of each IG is 690 V. This is converted via the two VSCs to DC and then back to a 690 V maximum, three-phase voltage that is stepped up to 33 kV by a transformer located in the turbine tower base. VSCs are bulky and costly for high-voltage and high-power applications. In fact, the power electronic switching devices that are the building blocks of the VSCs in the Walney system are not commercially

© Springer Nature Switzerland AG 2020
O. Beik, A. S. Al-Adsani, *DC Wind Generation Systems*,
https://doi.org/10.1007/978-3-030-39346-5_3

Fig. 3.1 Walney offshore wind farm [2]

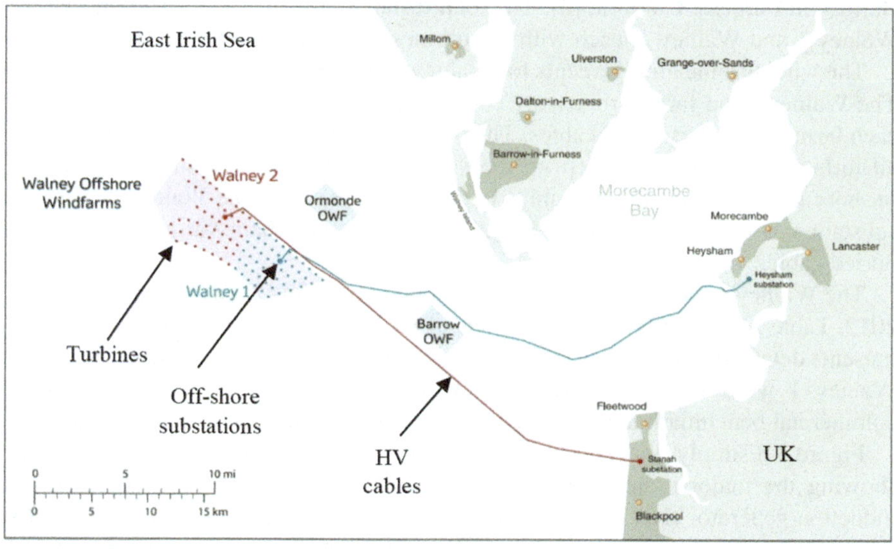

Fig. 3.2 Walney wind farm turbine arrangements [2]

available to date at high voltages; hence, high-voltage, high-power VSCs necessitate series of connected, modular, or multilevel structures to realize the voltage requirements and a number of parallel units to achieve the required power levels. This is one factor that constrains the wind industry to low-voltage commercial industrial VSC designs and the necessary step-up transformer to realize reasonable transmission

Table 3.1 Walney wind farm summary of details [2]

Item	Walney 1	Walney 2
Wind turbine type	Siemens SWT-3.6-107	Siemens SWT-3.6-120
Number of turbines	51	51
Output power (MW)	183.6	183.6
Transmission type	MVAC/HVAC	
Operating voltage level (kV)	33/132	
Offshore substation (MVA)	2×120	2×120
Distance between turbines (m)	749–958	
Distance from sub. to shore (km)	44.4	43.7
Approximate total area (km^2)	34	39
Approximate water depth (m)	21–26	

Table 3.2 Details of Siemens SWT-3.6-107 wind turbine [5]

Operating data	
Cut-in wind speed (m/s)	3–5
Nominal power at (m/s)	13–14
Cut-out wind speed (m/s)	25
Average wind speed at 84 m (m/s)	>9
Generator, converter, and transformer	
Generator type	IG
Generator rated power (MW)	3.6
Generator rated voltage (V)	690 (RMS line-to-line)
Generator number of poles	4
Frequency (Hz)	50
Generator rated speed (RPM)	1500
Power electronic converter	Full-scale back-to-back VSCs
Location of VSCs	One in nacelle, one in tower
Turbine transformer location	Tower base
Transformer voltage ratio (kV)	47.83 (=0.69/33)
Rotor	
Rotor speed (RPM)	5–13
Rotor diameter (m)	107
Blade length (m)	52
Rotor type	Three-blade horizontal axis
Rotor position	Upwind
Hub height	83.5 m
Power regulation	Full-span pitch regulation
Gearbox ratio	1:119
Gearbox type	Three-stage planetary/helical
Mechanical brake	Hydraulic disc brake

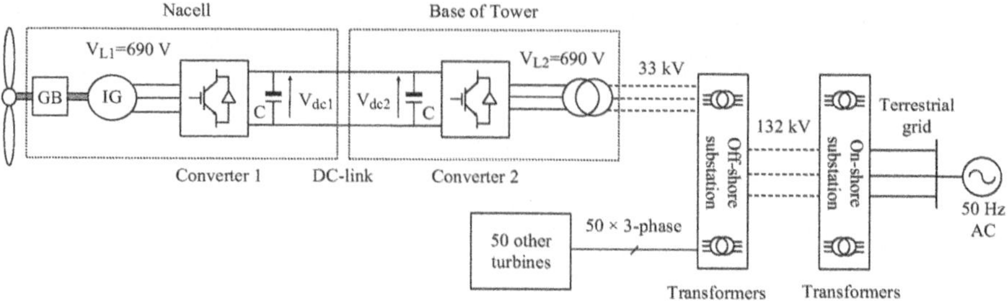

Fig. 3.3 Walney wind turbine connection to terrestrial grid

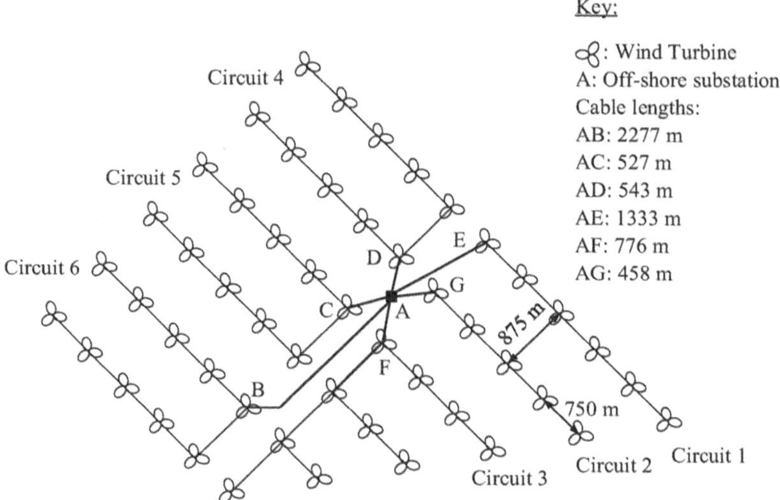

Fig. 3.4 Walney 1 site layout

voltage levels. Further, VSCs also have reliability issues in terms of passive capacitors sited on the DC link. These components are one of the major causes of failure in industrial VSC drive systems compromising wind turbine service life and necessitating maintenance/refit schedules. Given the low-voltage nature of the wind energy conversion, high currents have to be handled in the turbine tower; currents in the order of 3000 Amps for each 3.6 MW turbine is discussed later.

3.2   Walney System Component Model

The site layout for the Walney 1 wind farm is illustrated in Fig. 3.4. Walney 1 consists of six circuits. Circuit 1 and Circuit 2 have six and five wind turbines, respectively, whereas the other circuits have ten turbines. The electrical single-line

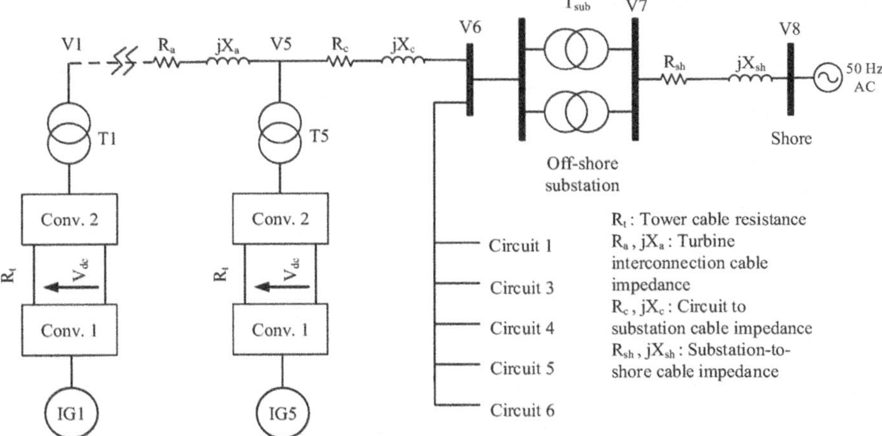

Fig. 3.5 Walney 1 site layout

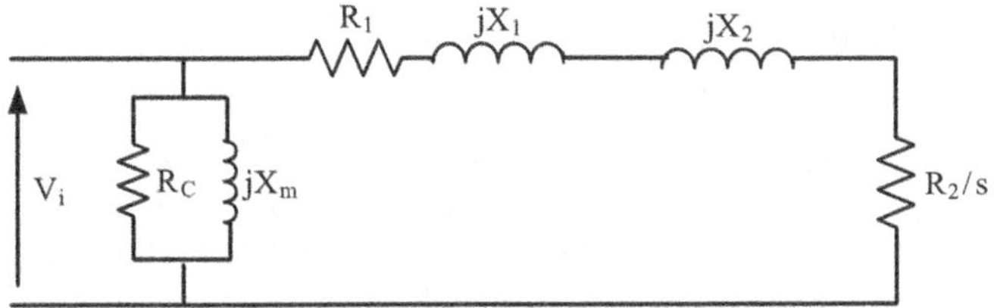

Fig. 3.6 Balanced, per-phase, steady-state equivalent circuit IG model

diagram of Walney 1 is illustrated in Fig. 3.5 showing the detailed components of Circuit 2. Each turbine is equipped with an induction generator, two back-to-back VSCs, Converter 1 and Converter 2, and a transformer. The IG and Converter 1, Fig. 3.5, are located in the nacelle, while Converter 2 and transformer are at the bottom of turbine tower. In this section, models for all the component are presented.

3.2.1   Induction Generator (IG) Model

The turbine generators used in Walney 1 are squirrel cage IG manufactured by Siemens, the conventional, balanced per-phase, steady-state equivalent circuit model for which is as shown in Fig. 3.6 and parameters listed in Table 3.3.

Table 3.3 Induction generator (IG) parameters

Description	Parameter	Value (Ω)
Stator resistance	R_1	0.0019
Rotor resistance referred to stator	R_2	0.0019
Loss resistance	R_c	4.41
Magnetizing reactance	X_m	1.16
Stator leakage reactance	X_1	0.048
Rotor leakage reactance referred to stator	X_2	0.035

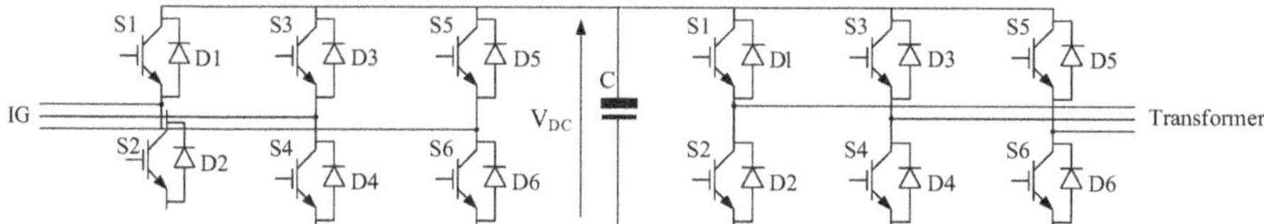

Fig. 3.7 VSC model

3.2.2 VSC Model

Referring to Fig. 3.5, the two power electronic VSCs decouple the variable frequency output of the generator from the fixed frequency grid. Converter 1 controls the speed of the generator to extract the maximum power from the wind via MPPT, while Converter 2 regulates the DC-link voltage to keep it constant. Converter 2 therefore maintains a constant DC-link voltage at all times, i.e., from full load to no load. Thus, all circuits from the DC link to the point of connection at the onshore grid are always excited to full nominal voltage. Pulse width modulation (PWM) vector control of the two VSCs is implemented to maximize power conversion of the induction generator and the DC link to AC output of Converter 2. Converter 1 is placed in the nacelle, while Converter 2 is located at the bottom of the turbine tower adjacent to the tower transformer. The output of Converter 2 is synchronized to the onshore 50 Hz grid. The back-to-back VSC model is presented in Fig. 3.7. Each converter consists of three-phase legs, and there are two power electronic switching devices, for example, S1 and S2, per leg. To allow higher power, each power electronic switch consists of a number of Insulated-Gate Bipolar Transistors (IGBTs). The antiparallel diodes, for example, D1 and D2, are freewheeling diodes that allow the flow of reverse energy when appropriate. The PWM signals for the IGBTs gates are generated by comparing a control (modulating) signal which is essentially normalized phase voltage of the AC terminals with a high-frequency sawtooth (triangular) carrier signal. Note, in the industry, these power electronic switching devices are often referred to as valves, a throwback to earlier valve technologies. In the steady-state system analysis presented later, the VSCs are modeled via their input, output, and efficiencies.

3.2.3   Transformer Model

Each turbine has an associated 4.5 MVA, 0.69/33 kV tower-mounted transformer. Further, the offshore substation has two transformers each rated at 120 MVA, 33/132 kV. The conventional, balanced, per-phase equivalent circuit used to model the turbine and the offshore substation transformers is shown in Fig. 3.8, and their parameters are listed in Table 3.4. Note that the output of Converter 2 is fixed at onshore electrical utility grid frequency of 50 Hz.

3.2.4   AC Cables Model

The inter-array cables connecting the turbines in Fig. 3.4 are composed of 3×150 mm^2 copper conductors while each circuit is connected to the offshore substation using a cable composed of 3×500 mm^2 copper conductors. The cable connecting the offshore substation to shore is 44.4 km in length and is composed of

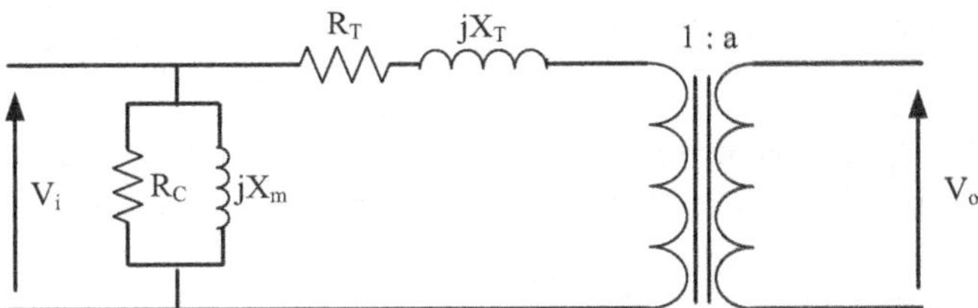

Fig. 3.8 Balanced, per-phase, steady-state transformer equivalent circuit model

Table 3.4 Transformer parameters at 50 Hz

Description	Parameter	Value Turbine transformer	Substation transformer
Rated power (MVA)	–	4.5	2×120
Rated voltage (kV)	–	0.69/33	33/132
No-load losses (kW)	–	3.7	63.3
Load losses (kW)	–	46.5	370
Percentage impedance (%)	–	6.4	10.5
Iron loss resistance (Ω)	R_c	128.68	1741.71
Magnetizing reactance (Ω)	X_m	10.62	213.5
Transformer equivalent resistance referred to primary (Ω)	R_T	0.001	0.002
Transformer equivalent leakage reactance referred to primary (Ω)	X_T	0.00677	0.089
Transformer ratio	a	$\frac{33}{0.69} = 47.83$	$\frac{132}{33} = 4$

Fig. 3.9 Cable-balanced, per-phase, steady-state, Pi equivalent circuit model

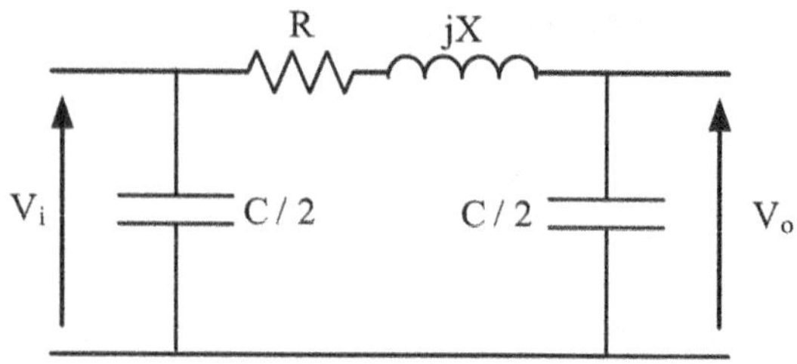

Table 3.5 Cable equivalent circuit parameters at 50 Hz

Description/parameters		Circuit 2		
		Inter-array between turbines	Inter-array to offshore substation	Offshore substation to onshore
Cable length (km)	–	0.75	0.458	44.4
Material	–	Copper		
Cable size (mm^2)	–	3×150	3×500	3×630
Rated voltage (kV)	–	33	33	132
AC resistance (Ω/km)	R	0.16	0.0515	0.028
Reactance (Ω/km)	X	0.116	0.0974	0.116
Capacitance (μF/km)	C	0.191	0.3	0.21

3×630 mm^2 copper conductors. The balanced, per-phase Pi (π) equivalent circuit model of the cables is shown in Fig. 3.9, and their parameters are listed in Table 3.5.

3.2.5 DC-Link Model

The connection between Converter 1 and Converter 2 is via a DC link. Information for the DC-link cable was not accessible to the author. Therefore, in order to calculate the DC-link cable impedance, the DC cable cross section of the DC cable is estimated using voltage and power information as follows. Considering the PWM switching scheme for the VSC, the DC-link voltage at the output of the Converter 1, Fig. 3.5, is given by:

$$V_{dc1} = \frac{2\sqrt{2}}{\sqrt{3}\, m_{a1}}\, V_{L1}(\text{V}) \qquad (3.1)$$

where

V_{L1} is the IG RMS line-to-line voltage fundamental component.
m_{a1} is amplitude modulation index for Converter 1.

The amplitude modulation index is defined as the ratio of amplitudes of control signal over carrier signal. Alternatively, the amplitude modulation index can be defined as the ratio of the IG operating speed over maximum speed where a simple *V/f* control scheme is implemented:

$$m_{a1} = \frac{\widehat{V}_{control}}{\widehat{V}_{tri}} = \frac{\omega_{gen}}{\omega_{max}} \tag{3.2}$$

where

$\widehat{V}_{control}$ is the amplitude of the control signal.
$\widehat{V}_{tri}$ is the amplitude of the carrier signal.
ω_{gen} is the IG operating speed.
ω_{max} is the IG maximum speed.

The DC-link current is calculated as follows:

$$I_{dc} = \frac{P_{dc1}}{V_{dc1}} \, (A) \tag{3.3}$$

where P_{dc1} is the input active power to the DC link from Converter 1.

The average current through the DC-link capacitor is zero; therefore, the DC current throughout the entire length of the DC link is the same. Assuming operation at the maximum IG output speed and voltage, the output DC-link voltage from Converter 1 is 1126.77 V. The two capacitors in the DC link support a constant DC-link voltage and reduce voltage ripple. Note that the IG is assumed to be closely connected to Converter 1 such that the resistances of the connecting inter-leads are negligible. Considering a maximum generated power output of each IG of 3.6 MW, the DC-link cable conductor cross-section area and resistance are thus calculated:

$$A_{dc} = \frac{I_{dc}}{J_{dc}} \, (mm^2) \tag{3.4}$$

$$R_{dc} = \frac{\rho_{cu} \, l_{dc}}{A_{dc}} \, (\Omega) \tag{3.5}$$

where

ρ_{cu} is the electrical resistivity of the DC-link cable.
l_{dc} is the DC-link cable length that runs from the nacelle to the bottom of the tower.
J_{dc} is the DC-link cable current density that is assumed to be 1.0 A/mm^2, i.e., the same as the AC cables current density in order to have the same thermal performance for the cables.

Table 3.6 DC-link calculations

Parameter	Description	Value
V_{L1}	IG RMS line voltage (V)	690
m_{a1}	Converter 1 modulation index	1.0
V_{dc1}	Converter 1 DC-link voltage (V)	1126.77
P_{dc1}	Input power to the DC link (MW)	3.6
I_{dc}	DC-link current (A)	3195
J_{dc}	DC-link cable current density (A/mm^2)	1.0
A_{dc}	DC-link cable cross-section area (mm^2)	3195
ρ_{cu}	Copper resistivity ($\mu\Omega$m)	0.0172
l_{dc}	DC cable length (m)	83.5
R_{dc}	DC-link cable resistance (mΩ)	0.45
V_{dc2}	Converter 2 DC-link voltage (V)	1123.9
ΔV_{dc}	DC-link voltage drop (V)	2.87
$P_{loss-dc}$	DC-link power loss (kW)	9.2

The power loss and voltage drop in the DC-link cable are calculated from:

$$P_{loss-dc} = 2R_{dc}\, I_{dc}^2 (\text{W}) \tag{3.6}$$

$$V_{dc2} = V_{dc1} - 2R_{dc}\, I_{dc} (\text{V}) \tag{3.7}$$

where the factor 2 is due to the sum of the resistances of the two equal cable conductors in the DC link, i.e., the go and return conductors. The DC-link calculation results are given in Table 3.6.

3.3   Walney System Analysis at Different Loads

The Walney 1 wind farm has been modeled using equivalent models for the induction generator, power electronic converters, DC-link cables, transformer, and cables as discussed previously. MATLAB programs have been written to perform the system power flow study. Different loading conditions were studied and analysis conducted with the following assumptions:

1. The distance between the induction generator and Converter 1 in the nacelle is negligible and hence neglected.
2. The distance between Converter 2 and the transformer at the bottom of the tower is also negligible and hence neglected.
3. Converter 2 is controlled to deliver a unity power factor output for all turbines and at all loading conditions, except for the no-load case.
4. Therefore, based on (3), all of the circuits connect to the offshore substation with the same power factor.

5. The length of the undersea transmission cable is the distance from the offshore substation to an immediate substation onshore. In other words, the distance that might exist from the shore termination station to the onshore substation is not considered.
6. The turbines produce no power at no load; hence, since the system is energized, the power to supply the system losses, for monitoring, for measuring devices, and for other ancillary services is taken from the onshore grid to which the wind farm is connected.

3.3.1 Full-Load Results

The results of the system analysis at full-load operation are presented in Tables 3.7 and 3.8. Here, full load assumes that all the IGs in the wind farm generate 3.6 MW to their respective Converter 1. Full-load condition may occur when the turbines are operating within a certain wind velocity range, i.e., between 13 and 25 (m/s) for Siemens SWT-3.6-107 wind turbine. The frequency of this loading condition is not considered here since the primary objective of the analysis is to deduce the system operating envelope.

Referring to the system schematic shown in Fig. 3.5 and the results of Tables 3.7 and 3.8, the following points are observed from the system full-load analysis:

1. The scalar voltage drop from the first turbine to the offshore substation input is 0.4% at full load, i.e.:

$$\left(\frac{V_1 - V_6}{V_1}\right) 100 = \left(\frac{32.74 - 32.61}{32.74}\right) 100 = 0.4\% \tag{3.8}$$

This is a considerably small drop voltage resulting in a small voltage difference at the output of each turbine.

2. The phase voltage angle decreases from the turbines to the offshore substation and from the offshore substation to the onshore substation, i.e.:

$$\begin{aligned} \text{angle } (V_1) > \text{ angle } (V_2) > \text{ angle } (V_3) > \text{ angle } (V_4) > \\ \text{angle } (V_5) > \text{ angle } (V_6) > \text{ angle } (V_7) > \text{ angle } (V_8) \end{aligned} \tag{3.9}$$

This results in an active power flow from Turbine 1 to the offshore substation and from the offshore substation to the onshore at full-load operation.

3. The power factor at the output of the turbine transformer is leading as defined by the measurements at the transformer terminals assuming transformer-to-shore power flow. Moreover, the power factor at the output bus bar of the offshore substation is leading, resulting in reactive power being absorbed by the offshore

Table 3.7 Walney 1, full-load analysis results: voltages and currents

Turbine					
Turbine no.	1	2	3	4	5
IG line voltage (V)	690	689.7	689.2	688.4	687.37
IG phase current (A)	3075.7	3076.9	3079.2	3082.7	3087.5
DC-link voltage at Converter 2 (V)	1123.9	1123.5	1122.6	1121.3	1119.6
DC-link current (A)	3052.65				
Converter 2 line voltage (V)	688.24	687.98	687.45	686.7	685.61
Converter 2 power factor	1.0				
Converter 2 phase current (A)	2835	2836	2838.2	2841.4	2845.8
Transformer output voltages and currents					
Phase voltage and angle (kV), (deg.)	18.90, −2.79	18.89, −2.80	18.88, −2.84	18.86, −2.89	18.83, −2.95
Transformer line voltage (kV)	32.74	32.73	32.71	32.67	32.62
Power factor (leading)	0.9981	0.9981	0.9980	0.9980	0.9979
Phase current (A)	59.22	59.24	59.28	59.35	59.45
Inputs at the offshore substation					
Phase voltage and angle (kV, degrees)				18.83, −2.99	
Input line voltage at the offshore substation (kV)				32.61	
Power factor				1.0	
Phase current from Circuit 2 (A)				296.30	
Outputs from the offshore substation					
Phase voltage and angle (kV, degrees)				75.34, −3.40	
Offshore substation line voltage (kV)				130.48	
Phase current (A)				751.46	
Power factor (leading)				0.998	
Inputs at the onshore substation					
Phase voltage and angle (kV, degrees)				74.19, −6.33	
Line voltage at the onshore substation (kV)				128.50	
Power factor (lagging)				0.99	
Phase current (A)				763.10	

Table 3.8 Walney 1, full-load analysis results: power and losses

Power and losses in the nacelle and tower					
Turbine power (kW)				3600	
IG output power (kW)				3492	
Converter 1 output power (kW)				3439.62	
Converter 2 output power (kW)				3379.39	
IG efficiency (%) and losses (kW)				97	108.00
Converter 1 efficiency (%) and losses (kW)				98.5	52.38
Converter 2 efficiency (%) and losses (kW)				98.5	51.46
DC-link power loss (kW)				8.77	
Turbine transformer efficiency (%)				99.18	
Turbine transformer loss					
Transformer no.	1	2	3	4	5
Loss (kW)	27.74	27.76	27.79	27.84	27.90
Turbine link cable					
From–to		1 to 2	2 to 3	3 to 4	4 to 5
Inter-array cable loss (kW)		1.26	5.05	11.36	20.21
Cable loss, Circuit 2 to the offshore substation (kW)					6.21
Input power and losses at offshore sub. from Circuit 2					
Power (kW)	16,713.83		Losses (kW)	1286.17	
Input power and losses at offshore sub. from Circuits 1–6					
Power (kW)	170,481.05		Losses (kW)	13,118.95	
Offshore substation transformers efficiency (%)				99.27	
Loss at offshore substation (kW)				1248.35	
Cable losses, offshore to onshore substations (kW)				2110.20	
Total power at onshore substation (kW)				167,346.54	
Total losses at onshore substation (kW)				16,477.50	
Total system efficiency (%)				91.15	

substation. The power factor at the onshore bus bar is lagging. Therefore, the line capacitance of the cable connecting the offshore substation to the onshore substation is providing some of the reactive compensation for the system at full-load operation.

4. Loss analysis shows that:

- The total loss for the Walney 1 system is 16,477.50 kW.
- Circuits 1–6 loss contributes to 79.62% of the total loss.
- The two offshore substation transformers contribute to 7.58% of the total loss.
- Losses in the cable connecting the offshore to onshore substations contribute to 12.80% of the total loss.
- Walney 1 efficiency is 91.15% at full load.

The offshore substation and the cable connecting the offshore substation to the onshore bus bar make up a large portion of the total system loss at full-load operation. Note that the result is specific to this benchmark system and is not to be a generic conclusion for other systems.

3.3.2 Results for Three-Quarter, Half-, and One-Quarter Loads

Power flow results for the Walney wind farm at three-quarter, half-, and one-quarter loads are presented in Tables 3.9, 3.10, 3.11, 3.12, 3.13, and 3.14. At three-quarter load, generator power is 2.7 MW while the generator produces 1.8 MW and 0.9 MW at half and quarter loads, respectively. Note that the PWM vector control of the IG by Converter 1 keeps a constant 97% efficiency for the machine as the load changes. Also, the efficiency of the VSCs for other loads is assumed to be same as full load, i.e., 98.5%.

The voltage drops from turbine one to the offshore substation at different loads are:

$$1.\ \text{Three-quarter load}: \left(\frac{V_1 - V_6}{V_1}\right) 100 = \left(\frac{32.78 - 32.68}{32.78}\right) 100$$

$$= 0.31\%. \tag{3.10}$$

$$2.\ \text{Half load}: \left(\frac{V_1 - V_6}{V_1}\right) 100 = \left(\frac{32.83 - 32.76}{32.83}\right) 100 = 0.21\%. \tag{3.11}$$

$$3.\ \text{One-quarter load}: \left(\frac{V_1 - V_6}{V_1}\right) 100 = \left(\frac{32.88 - 32.85}{32.88}\right) 100 = 0.1\%. \tag{3.12}$$

Therefore, as the load decreases, the voltage drop from the turbines to the offshore substation reduces. Hence, as the load decreases, the system nodal voltages increase. For all loads, the absolute value of the voltage angle decreases from turbine to the onshore substation.

For all loads, the IGs receive reactive power. Transformer reactive power is offset by the interconnection line capacitance. The power factor at the offshore substation input is unity for all loads; however, the offshore substation output power factor is leading. Therefore, for all loads, reactive power for the offshore substation transformers is being supplied by the transmission line capacitance connecting the offshore to onshore substation. Nonetheless, the power factor at the onshore substation is lagging which means the wind farm injects reactive power to the grid.

Table 3.9 Walney 1, three-quarter-load analysis results: voltages and currents

Turbine					
Turbine no.	1	2	3	4	5
IG line voltage (V)	690	689.80	689.41	688.82	688.04
IG phase current (A)	2306.76	2307.41	2308.72	2310.70	2313.33
DC-link voltage at Converter 2 (V)	1123.89	1123.57	1122.93	1121.97	1120.69
DC-link current (A)	2289.49				
Converter 2 line voltage (V)	688.24	688.05	687.65	687.07	686.28
Converter 2 power factor	1.0				
Converter 2 phase current (A)	2126.17	2126.77	2127.99	2129.81	2132.25
Transformer output voltages and currents					
Phase voltage and angle (kV), (deg.)	18.93, −2.09	18.92, −2.10	18.91, −2.12	18.90, −2.16	18.88, −2.21
Transformer line voltage (kV)	32.78	32.77	32.76	32.73	32.69
Power factor	1				
Phase current (A)	44.40	44.41	44.44	44.48	44.53
Inputs at the offshore substation					
Phase voltage and angle (kV, degrees)				18.87, −2.24	
Input line voltage at the offshore substation (kV)				32.68	
Power factor				1.0	
Phase current from Circuit 2 (A)				222.06	
Outputs from the offshore substation					
Phase voltage and angle (kV, degrees)				75.51, −2.54	
Offshore substation line voltage (kV)				130.78	
Phase current (A)				562.61	
Power factor (leading)				0.997	
Inputs at the onshore substation					
Phase voltage and angle (kV, degrees)				74.54, −4.71	
Line voltage at the onshore substation (kV)				129.11	
Power factor (lagging)				0.97	
Phase current (A)				582.34	

Table 3.10 Walney 1, three-quarter-load analysis results: power and losses

Power and losses in the nacelle and tower					
Turbine power (kW)				2700	
IG output power (kW)				2619	
Converter 1 output power (kW)				2579.72	
Converter 2 output power (kW)				2534.54	
IG efficiency (%) and losses (kW)				97	81.00
Converter 1 efficiency (%) and losses (kW)				98.5	39.29
Converter 2 efficiency (%) and losses (kW)				98.5	38.60
DC-link power loss (kW)				6.58	
Turbine transformer efficiency (%)				99.32	
Turbine transformer loss					
Transformer no.	1	2	3	4	5
Loss (kW)	17.21	17.21	17.22	17.24	17.27
Turbine link cable					
From–to		1 to 2	2 to 3	3 to 4	4 to 5
Inter-array cable loss (kW)		0.71	2.84	6.38	11.35
Cable loss, Circuit 2 to the offshore substation (kW)					3.49
Input power and losses at offshore sub. from Circuit 2					
Power (kW)	12,561.79		Losses (kW)	938.21	
Input power and losses at offshore sub. from Circuits 1–6					
Power (kW)	128,130.21		Losses (kW)	9569.79	
Offshore substation transformers efficiency (%)				99.03	
Loss at offshore substation (kW)				1241.92	
Cable losses, offshore to onshore substations (kW)				1186.61	
Total power at onshore substation (kW)				125,792.79	
Total losses at onshore substation (kW)				11,998.32	
Total system efficiency (%)				91.35	

The total system efficiency for different loads is:

$$1. \text{ Three-quarter load} : 91.35\%. \tag{3.13}$$

$$2. \text{ Half load} : 91.31\%. \tag{3.14}$$

$$3. \text{ One-quarter load} : 90.23\%. \tag{3.15}$$

The system efficiency reduces slightly as the load is decreased. Hence, similar benefits and system performance are seen at all loads. However, note that the IG and VSC efficiencies are kept constant at all loads.

Table 3.11 Walney 1, half-load analysis results: voltages and currents

Turbine					
Turbine no.	1	2	3	4	5
IG line voltage (V)	690	689.87	689.61	689.22	688.70
IG phase current (A)	1537.84	1538.13	1538.70	1539.57	1540.74
DC-link voltage at Converter 2 (V)	1123.89	1123.68	1123.26	1122.62	1121.77
DC-link current (A)	1526.32				
Converter 2 line voltage (V)	688.24	688.11	687.85	687.46	686.94
Converter 2 power factor	1.0				
Converter 2 phase current (A)	1417.45	1417.71	1418.25	1419.05	1420.12
Transformer output voltages and currents					
Phase voltage and angle (kV), (deg.)	18.95, -1.39	18.95, -1.40	18.94, -1.41	18.93, -1.44	18.92, -1.47
Transformer line voltage (kV)	32.83	32.82	32.81	32.79	32.77
Power factor	1				
Phase current (A)	29.58	29.59	29.60	29.62	29.64
Inputs at the offshore substation					
Phase voltage and angle (kV, degrees)	18.92, -1.49				
Input line voltage at the offshore substation (kV)	32.76				
Power factor	1.0				
Phase current from Circuit 2 (A)	147.87				
Outputs from the offshore substation					
Phase voltage and angle (kV, degrees)	75.69, -1.69				
Offshore substation line voltage (kV)	131.10				
Phase current (A)	374.37				
Power factor (leading)	0.99				
Inputs at the onshore substation					
Phase voltage and angle (kV, degrees)	74.92, -3.10				
Line voltage at the onshore substation (kV)	129.76				
Power factor (lagging)	0.91				
Phase current (A)	407.84				

Table 3.12 Walney 1, half-load analysis results: power and losses

Power and losses in the nacelle and tower					
Turbine power (kW)				1800	
IG output power (kW)				1746.00	
Converter 1 output power (kW)				1719.81	
Converter 2 output power (kW)				1689.69	
IG efficiency (%) and losses (kW)				97	54.00
Converter 1 efficiency (%) and losses (kW)				98.5	26.19
Converter 2 efficiency (%) and losses (kW)				98.5	25.73
DC-link power loss (kW)				4.38	
Turbine transformer efficiency (%)				99.43	
Turbine transformer loss					
Transformer no.	1	2	3	4	5
Loss (kW)	9.69	9.69	9.69	9.69	9.70
Turbine link cable					
From–to		1 to 2	2 to 3	3 to 4	4 to 5
Inter-array cable loss (kW)		0.31	1.26	2.83	5.04
Cable loss, Circuit 2 to the offshore substation (kW)					1.55
Input power and losses at offshore sub. from Circuit 2					
Power (kW)	8389.03		Losses (kW)	610.97	
Input power and losses at offshore sub. from Circuits 1–6					
Power (kW)	85,568.13		Losses (kW)	6231.87	
Offshore substation transformers efficiency (%)				98.55	
Loss at offshore substation (kW)				1239.36	
Cable losses, offshore to onshore substations (kW)				530.17	
Total power at onshore substation (kW)				83,822.82	
Total losses at onshore substation (kW)				8001.40	
Total system efficiency (%)				91.31	

3.4   DC Offshore Wind Generation System

The preceding section analyzed the Walney benchmark system against which the DC wind generation scheme is compared. The DC wind generation scheme comprises of a high-voltage HG, passive rectification, DC/DC converters, HVDC interconnection, and HVDC transmission systems. A schematic view of the DC system based on the Walney wind farm is illustrated in Fig. 3.10.

The DC wind generation scheme potentially results in reduced system cost due to reduced component count and complexity. Compared to the Walney system energy conversion components, Fig. 3.3, the HG and passive rectifier in the DC system replace the IG, back-to-back VSCs, and tower-mounted transformer, yielding a much simplified hardware scheme and comprising of no active electronics and electrolyte DC-link capacitor. The HG output rectifier is mounted close or integral to the HG machine assembly, i.e., in the machine housing. To minimize passive

Table 3.13 Walney 1, one-quarter-load analysis results: voltages and currents

Turbine					
Turbine no.	1	2	3	4	5
IG line voltage (V)	690	689.94	689.81	689.62	689.36
IG phase current (A)	768.92	768.99	769.13	769.35	769.64
DC-link voltage at Converter 2 (V)	1123.89	1123.79	1123.58	1123.27	1122.85
DC-link current (A)	763.16				
Converter 2 line voltage (V)	688.24	688.18	688.05	687.86	687.60
Converter 2 power factor	1.0				
Converter 2 phase current (A)	708.72	708.79	708.92	709.12	709.39
Transformer output voltages and currents					
Phase voltage and angle (kV), (deg.)	18.98, −0.69	18.98, −0.70	18.98, −0.71	18.97, −0.72	18.97, −0.73
Transformer line voltage (kV)	32.88	32.88	32.87	32.86	32.85
Power factor	1				
Phase current (A)	14.77	14.78	14.78	14.78	14.79
Inputs at the offshore substation					
Phase voltage and angle (kV, degrees)				18.96, −0.74	
Input line voltage at the offshore substation (kV)				32.85	
Power factor				1.0	
Phase current from Circuit 2 (A)				73.75	
Outputs from the offshore substation					
Phase voltage and angle (kV, degrees)				75.89, −0.84	
Offshore substation line voltage (kV)				131.44	
Phase current (A)				188.13	
Power factor (leading)				0.97	
Inputs at the onshore substation					
Phase voltage and angle (kV, degrees)				75.33, −1.50	
Line voltage at the onshore substation (kV)				130.47	
Power factor (lagging)				0.73	
Phase current (A)				252.59	

Table 3.14 Walney 1, one-quarter-load analysis results: power and losses

Power and losses in the nacelle and tower					
Turbine power (kW)				900	
IG output power (kW)				873.00	
Converter 1 output power (kW)				859.90	
Converter 2 output power (kW)				844.85	
IG efficiency (%) and losses (kW)				97	27.00
Converter 1 efficiency (%) and losses (kW)				98.5	13.10
Converter 2 efficiency (%) and losses (kW)				98.5	12.87
DC-link power loss (kW)				2.19	
Turbine transformer efficiency (%)				99.39	
Turbine transformer loss					
Transformer no.	1	2	3	4	5
Loss (kW)	5.18	5.18	5.18	5.18	5.18
Turbine link cable					
From–to		1 to 2	2 to 3	3 to 4	4 to 5
Inter-array cable loss (kW)		0.08	0.31	0.70	1.25
Cable loss, Circuit 2 to the offshore substation (kW)					0.38
Input power and losses at offshore sub. from Circuit 2					
Power (kW)	4195.61		Losses (kW)	304.39	
Input power and losses at offshore sub. from Circuits 1–6					
Power (kW)	42,795.27		Losses (kW)	3104.73	
Offshore substation transformers efficiency (%)				97.10	
Loss at offshore substation (kW)				1240.68	
Cable losses, offshore to onshore substations (kW)				140.31	
Total power at onshore substation (kW)				41,415.46	
Total losses at onshore substation (kW)				4485.71	
Total system efficiency (%)				90.23	

components and improve generated DC power quality, a nine-phase winding design
is chosen for the HG. For the same ampere-turns per phase and total copper losses,
the nine-phase winding yields a 4.2% higher DC-link voltage (and hence power)
when compared to a three-phase winding solution due to an improved winding
distribution factor. The peak-peak to average DC-link voltage ripple is also reduced
from 14% for the three-phase design to 1.53% for the nine-phase design, effectively
negating the requirement for DC-link smoothing capacitance. This will be discussed
in detail in Chap. 5.

3.4.1 Hybrid Generator (HG)

A brief discussion of the HG concept is presented in this section. However, the
detailed design and analysis of the HG are addressed in Chap. 5. The HG is a

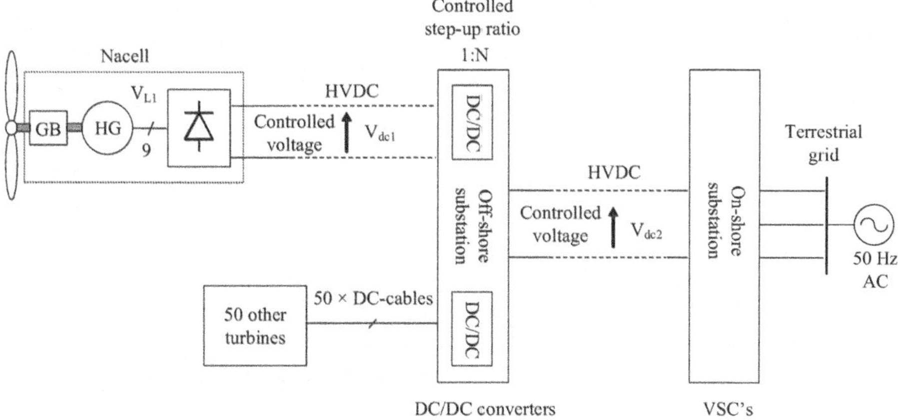

Fig. 3.10 DC wind generation scheme with high-voltage hybrid generator (HG), HVDC interconnection, and HVDC transmission

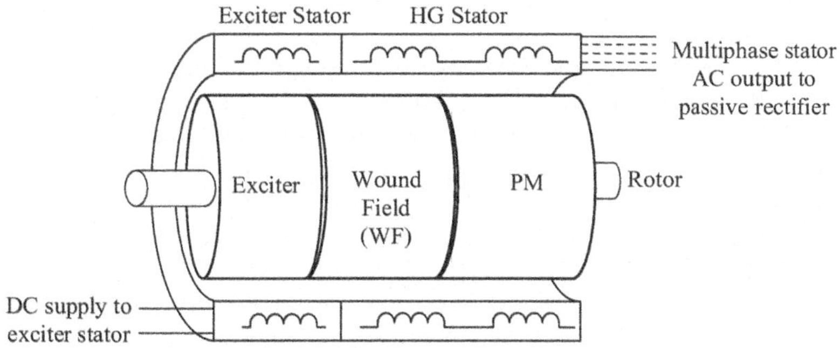

Fig. 3.11 Schematic view of the HG

multiphase, high-voltage generator that uses two rotor sections, namely, a PM rotor and a WF rotor, to provide the total machine excitation. Hence, the name hybrid refers to the combination of these two rotors excitation schemes. The PM and WF rotor sections exist on one rotor assembly inside one machine housing. Thus, the rotors rotate with the same speed. A schematic view of the HG is depicted in Fig. 3.11. DC current for the WF rotor excitation is provided via a brushless exciter system common to industrial synchronous machine systems.

Therefore, the HG combines the output voltage due to a fixed field from the PM rotor and a controlled variable voltage due to the variable field of the WF rotor. The total machine output voltage is the sum of the voltages due to PM and WF sections. The WF therefore contributes a controllable variable voltage at the stator output over a limited range. The range of the total voltage variation depends on the design of the WF rotor of the HG. Figure 3.12 shows the HG in a wind turbine system.

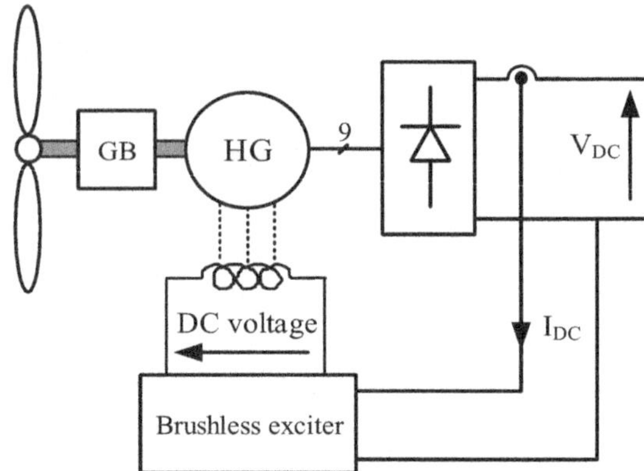

Fig. 3.12 HG in a wind turbine

3.4.2 Gearbox

The gearbox used in the benchmark (Walney) wind turbines is a Siemens-manufactured three-stage planetary/helical gearbox with a ratio of 1:119. Figure 3.13 illustrates example planetary and helical gear designs. Since the nominal turbine rotor speed is 13 RPM for the Siemens wind turbine, this input to the low-speed gearbox shaft is increased to around 1500 RPM on the high-speed gearbox output shaft to drive the IG. This speed is defined as the IG nominal speed. Other gearbox ratios with lower stages have been proposed in the literature. Polinder [43] considers different generators and gearboxes including a doubly-fed induction generator with a three-stage and single-stage gearbox having nominal rated speeds with 1200 RPM and 90 RPM rates speed, respectively, a direct drive synchronous and a PM generator with 15 RPM rated speed and a PM generator with single-stage gearbox with a rated speed of 90 RPM. In [43], the ratio of single-stage and two-stage gearbox is 6 and 80, respectively. A comprehensive study of other gearbox stages and ratios is discussed in detail in [54]. However, here, a two-stage gearbox with a ratio of 46.154 is chosen to increase the turbine rotor speed of 13 RPM to a rated speed of 600 RPM for the HG. This ratio is chosen such that the HG rated speed is consistent with a wound field SG chosen to benchmark the HG electromagnetic design, as discussed in Chap. 4.

3.4.3 Voltage Control Scenarios

Although the Walney IGs are generating at a variable voltage, the remaining electrical system is required to be operated at constant voltage, i.e., at around the nominal rated voltage at all times and for all wind speed conditions. This is to ensure operation of the back-to-back VSCs and ancillary components in the system including monitoring, measurements, and maintenance equipment. For the DC scheme, the

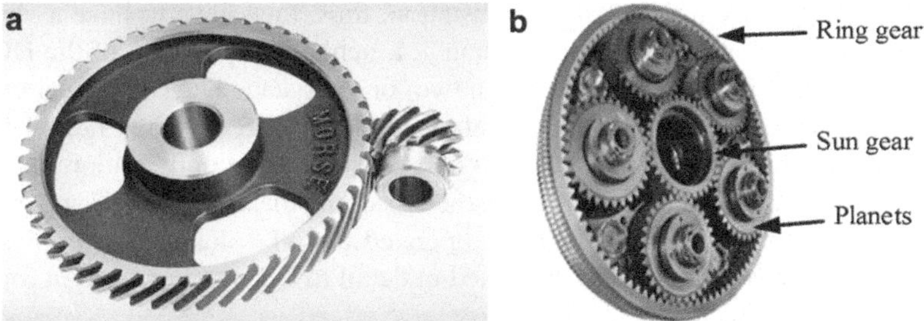

Fig. 3.13 Example planetary and helical gearbox implementation [55]. (**a**) Helical gears (**b**) Planetary gears

system voltage is allowed to vary. Two control methods could be considered for the HG output DC-link voltage:

1. Fixed voltage
2. Variable (but controlled) voltage

1. Fixed voltage. In the fixed voltage approach, the rectified output voltage of the HG is kept around its nominal DC value at all wind speed conditions, i.e., from cut-in to cut-out velocities. This is implemented by controlling the HG WF current. In this case, the voltage transfer ratio of the DC/DC converters, Fig. 3.10, is fixed at 1:4, chosen to be the same as the Walney wind farm turbine transformer step-up ratio (i.e., 33/132), a ratio that is easily implementable by DC/DC converter circuits. Note that there will be small differences in voltage transfer ratios among the system DC/DC converters to maintain power flow and avoid rectifier turnoff due to variations in the wind farm turbine speeds (and hence generated DC voltage perturbations). The DC/DC converter voltage transfer ratio is realized by pulse width modulation (PWM) of the converter power electronic devices. This control function produces a switching duty ratio "D" that is directly related to the voltage transfer ratio of the DC/DC converters:

$$\text{Voltage transfer ratio} = n_t D \frac{V_{\text{input}}}{V_{\text{output}}} \tag{3.16}$$

where n_t is the turns (transfer) ratio of the DC/DC isolation transformer.

2. Variable but controlled voltage. In the second control scheme, the rectified output voltage of the HG system varies with wind velocity below the rated velocity (refer to Fig. 2.11). Above the rated velocity, the WF is controlled to keep the DC link at nominal maximum, although again, there will be small perturbations around the maximum value due to wind transients. Therefore, the voltage across the interconnection cables between the wind turbines and the offshore substation is variable but controlled. However, the voltage across the transmission cables connecting the offshore to onshore substation is controlled to be fixed via the

DC/DC converters at the offshore substation; thus, they have to have a variable voltage transfer ratio that is greater than 1:4, achieved by choice of the DC/DC isolation transformer turns ratio or by a two- or multistage DC/DC converter. The variable voltage control scenario with variable voltage transfer ratio DC/DC converter at the offshore substation is chosen since this configuration results in a lower WF current requirement and hence machine mass for the HG and ensures that the offshore to onshore cable is operated at full voltage, thus minimizing ohmic loss. These features are discussed in detail in Chap. 5 but brought forward here as input to the system.

3.4.4 DC/DC Converter

Different topologies have been proposed for DC/DC converters at high voltage and high power. The single active bridge (SAB) DC/DC converter topology uses an input VSC, an intermediate transformer, and an output rectifier, as shown in Fig. 3.14.

The input VSC converts DC to an AC voltage with medium or high frequency where this voltage is stepped up by the intermediate transformer and then rectified by a passive rectifier. The topology shown in Fig. 3.14 is a three-phase SAB. Single-phase SAB topologies can also be realized by a single-phase VSC, single-phase transformer and rectifier. However, the higher the number of phases improves the volume power processing capability and can improve DC quality if interleaving technologies are implemented. Power and frequency control are realized using the active power electronic switches in this topology.

If the output rectifier in the SAB is replaced with a VSC, a dual active bridge (DAB) topology is realized for the DC/DC converter as shown in Fig. 3.15. Similar to the SAB, a single-phase DAB can also be implemented.

Jovcic [56, 57] has proposed a high-power DC/DC converter topology with high step-up ratio DC/DC converter that utilizes thyristors and LC circuits. The DC/DC converter does not include a step-up transformer, as shown in Fig. 3.16. Jovcic has shown that this proposed topology can achieve efficiencies of 95% for a 5 MW DC/DC converter while stepping up 4 kV to 80 kV, i.e., gain of 20.

The DC/DC converters are modeled as a component with specific parameters and efficiency characteristics.

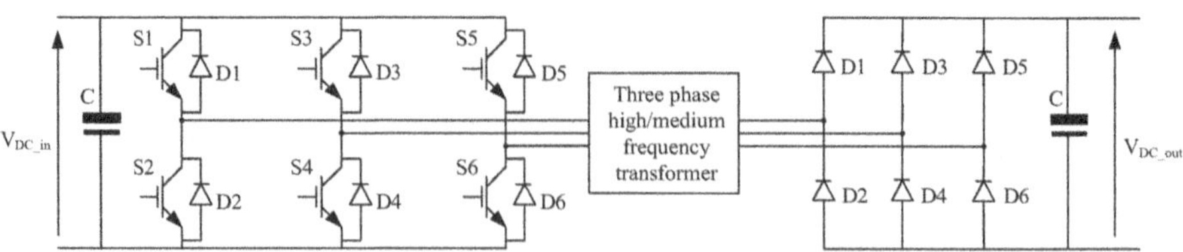

Fig. 3.14 Three-phase, single active bridge (SAB) topology for DC/DC converter

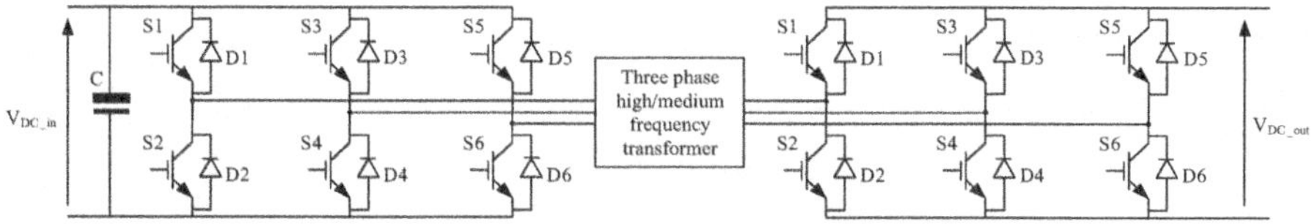

Fig. 3.15 Three-phase dual active bridge (DAB) topology for DC/DC converter

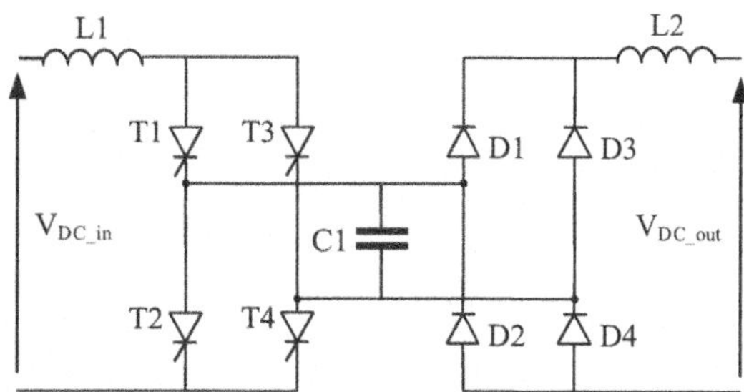

Fig. 3.16 LC circuit-based DC/DC converter with thyristors

3.4.5 Turbine Safety Considerations

For the Walney wind farm generators, variable power-speed operation of the wind turbine is implemented by the back-to-back VSCs as well as mechanical controls such as blade pitch angle control. In the DC system, the WF voltage and mechanical blade pitch control are the main mechanisms for generator output power-speed control. This results in component reduction, which in turn results in lower power train mass. As with the Walney scheme, wind turbine control above cut-out speed would be mainly performed by mechanical control mechanisms, i.e., pitch and yaw control. Once the cut-out velocity is exceeded, the WF current in the HG is reduced to zero, the turbine blades pitched toward feather position, and/or the nacelle is moved out of the wind to limit the overspeed of the turbine. Moreover, a mechanical brake is provided on the shaft as an ultimate mechanical control in both Walney and the DC scheme.

3.5 DC System Analysis

The DC system is analyzed and compared with the Walney wind farm, keeping the geometrical topology of the wind farm, i.e., distances, site layout, and turbine arrangements, unchanged. The single-line diagram of the DC system is shown in Fig. 3.17. The analysis studies are carried out with following assumptions:

1. The passive rectifier is integral to the generator and installed in the nacelle; hence, the distance between the HG and rectifier is neglected.

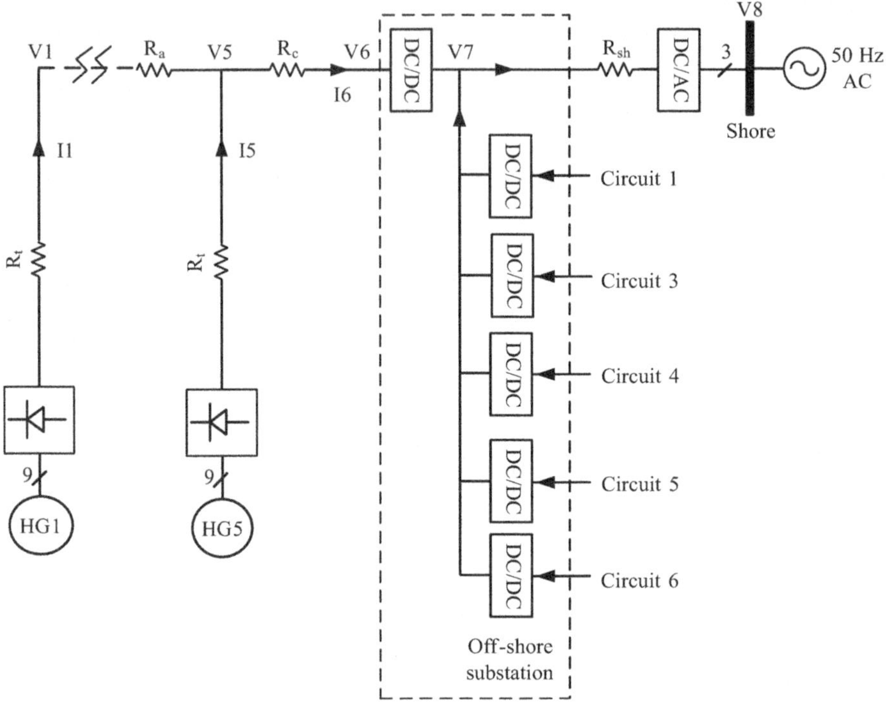

Fig. 3.17 Schematic of DC system

2. At full load, the voltage is at its rated value, and the voltage transfer ratio of the offshore substation DC/DC converters is nominally 1:4. As the wind velocity changes and the voltage drops, the DC/DC voltage transfer ratio increases.
3. The nominal output power of each turbine in the DC system is the same as that of the Walney system, i.e., 3.6 MW. Therefore, the HG and passive rectifier in the DC system are rated at 3.6 MW.
4. The rated RMS line-to-line voltage of the IG in Walney system is 690 V. However, the wind farm interconnection cables' voltage and hence the turbine transformer output voltage are usually rated for RMS line-to-line voltages up to 66 kV (RMS phase voltage 38.1 kV). The turbine transformer is eliminated in the DC system, and the nine-phase HG is rated at 38.1 kV RMS per phase (quoted in per phase since generator line-to-line voltages are not all equal).

Considering the nine-phase HG connected to a passive rectifier, the DC voltage at the rectifier terminals is obtained via:

$$V_{\text{dc}} = \frac{2\,m\,\sin\left(\pi/m\right)}{\pi}\,\widehat{V}_{\text{ph}}\ (\text{V}) \tag{3.17}$$

where

m is the number of phases.

$\widehat{V}_{\mathrm{ph}}$ is the peak phase voltage.

Therefore, for a 38.1 RMS kV phase voltage and nine-phase machine, the resultant DC output voltage is:

$$V_{\mathrm{dc}} = 105.6 \ (\mathrm{kV}) \tag{3.18}$$

3.5.1   DC Cable Calculations

In order to calculate the required cable cross-sectional areas and cable resistances, the cable current densities are assumed to be the same as the corresponding cables in Walney 1 wind farm at their highest loading; hence, skin effects are neglected in order to calculate a conservative estimate. This ensures the same copper thermal performance for both the AC and DC cables. Since the wind farm topology is unchanged, the length of the cables in the DC system is the same as the AC cable lengths in the Walney 1 wind farm. In Walney, the highest loaded inter-array cable is the one that connects Turbine 4 to Turbine 5, Fig. 3.15. The current in this cable at full load is the sum of currents from Turbines 1 to 4. Using the calculated currents stated in Table 3.7 and cross-section area of the AC cables listed in Table 3.5, the AC cable current density is deduced to be:

$$J_{\mathrm{ac}}^{\mathrm{array}} = \frac{I_{\mathrm{ac}}^{\mathrm{array}}}{A_{\mathrm{ac}}^{\mathrm{array}}} = \frac{273.09}{150} = 1.58 \ (\mathrm{A/mm^2}) \tag{3.19}$$

where

$I_{\mathrm{ac}}^{\mathrm{array}}$: Inter-array cable current
$A_{\mathrm{ac}}^{\mathrm{array}}$: Inter-array cable cross-section area

The inter-array DC current for the highest loaded cable in Fig. 3.17 is again the sum of currents from Turbines 1 to 4. Individual turbine current is calculated using turbine output power and rectified voltage:

$$I_{\mathrm{dc}}^{\mathrm{turbine}} = \left(\frac{P_{\mathrm{turbine}}}{V_{\mathrm{dc}}}\right) = \left(\frac{3.6 \times 10^6}{105.6 \times 10^3}\right) = 34.1 (\mathrm{A}) \tag{3.20}$$

Hence:

$$I_{\mathrm{dc}}^{\mathrm{array}} = 4 \ I_{\mathrm{dc}}^{\mathrm{turbine}} = 4 \ (34.1) = 136.36 \ (\mathrm{A}) \tag{3.21}$$

where

$I_{dc}^{turbine}$ is the turbine output current.

$P_{turbine}$ is the turbine power.

Therefore, the inter-array cable cross-sectional area is:

$$A_{dc}^{array} = \frac{I_{dc}^{array}}{J_{ac}^{array}} = \frac{136.36}{1.58} = 86.31 \ \left(mm^2\right) \tag{3.22}$$

Current density for the tower cable is considered to be the same as the one for inter-array cable. Therefore, the tower cable cross-sectional area is:

$$A_{dc}^{tower} = \frac{I_{dc}^{turbine}}{J_{ac}^{array}} = \frac{34.1}{1.58} = 21.58 \ \left(A/mm^2\right) \tag{3.23}$$

The current in the cable connecting Circuit 2 to the offshore substation is the sum of currents in Turbines 1 to 5. Using current values stated in Table 3.7 and cross-section area quoted in Table 3.5:

$$J_{ac}^{circuit} = \frac{I_{ac}^{circuit}}{A_{ac}^{circuit}} = \frac{296.54}{500} = 0.6 \ \left(A/mm^2\right) \tag{3.24}$$

Therefore, the cross-sectional area of the DC cable connecting Circuit 2 to the offshore substation in Fig. 3.17 is:

$$A_{dc}^{circuit} = \frac{I_{dc}^{circuit}}{J_{ac}^{circuit}} = \frac{5(34.1)}{0.6} = 284.2 \ \left(mm^2\right) \tag{3.25}$$

where

$I_{ac}^{circuit}$ is the current in the cable connecting Circuit 2 to the offshore substation.

$A_{ac}^{circuit}$ is the cross-section area for the cable connecting Circuit 2 to the offshore substation.

Similarly, for the transmission cable connecting offshore to onshore substation, the current density using Tables 3.7 and 3.5 is:

$$J_{ac}^{trans} = \frac{I_{ac}^{trans}}{A_{ac}^{trans}} = \frac{715.46}{630} = 1.2 \ \left(A/mm^2\right) \tag{3.26}$$

Thus, the current in the cable connecting offshore to onshore substation for the DC system is the sum of currents from 51 turbines and with consideration to a step-up ratio of 4 for the DC/DC converters. Therefore, the cross-sectional area of the corresponding DC cable for the offshore to onshore transmission cable is:

$$A_{dc}^{trans} = \frac{I_{dc}^{trans}}{J_{ac}^{trans}} = \frac{51(34.1)(1/4)}{1.2} = 362.31 \ \left(mm^2\right) \tag{3.27}$$

where I_{dc}^{trans} is the current in the cable connecting offshore to onshore substation.

The cable resistance for each DC cable is calculated using Eq. (3.5). Note that the total length of a DC cable is double the length of the point-to-point connection since there are two DC wires for each cable connection. A summary of DC cable calculations is listed in Table 3.15.

3.5.2   Results for Full-, Three-Quarter, Half-, and One-Quarter Loads

The analysis results for the DC system analysis at full load are presented in Table 3.16. At full load, all generators in the system generate 3.6 MW to their respective rectifier. As mentioned previously, as the wind velocity changes, the output voltage of the turbine HG changes. For the Siemens SWT-3.6-107 wind turbine, the turbine rotor speed changes from 5 RPM at cut-in to 13 RPM at rated speed. The output voltage of the HG changes linearly with turbine rotor speed. Moreover, as discussed in Chap. 2, turbine power is a cubic function of the turbine rotor speed. Therefore, the output voltage of the HG changes with cubic root of power. As discussed previously, the voltage across interconnection cables varies, but the offshore substation output voltage is constant. Therefore, the DC/DC converter step-up ratio varies as the wind velocity changes. Table 3.17 presents rotor speed, HG output voltage, and DC/DC converter ratio at different loads.

Table 3.15 DC system cables specifications

Description	Tower	Circuit 2		Offshore substation to onshore
		Inter-array between turbines	Inter-array to offshore substation	
Cable length (km)	0.0835	0.750	0.458	44.4
Material	Copper			
Copper resistivity ($\mu\Omega$m)	0.0172			
Rated voltage (kVDC)	110	110	110	425
Current density (A/mm^2)	1.58	1.58	0.6	1.2
Cross-section area (mm^2)	21.58	86.31	284.2	362.31
DC cables resistance (Ω)	0.13	0.3	0.055	4.2

Table 3.16 DC system full-load analysis

Turbine					
Turbine no.	1	2	3	4	5
HG phase voltage (kV)	38.11	38.10	38.09	38.08	38.08
HG DC voltage (kV)	105.6	105.59	105.57	105.54	105.54
HG DC current output (A)	32.90	32.91	32.91	32.92	32.92
DC voltage at the tower base, kV	105.6	105.59	105.57	105.54	105.50
Interconnection and transmission					
Input DC voltage at the offshore substation (kV)				105.49	
Input DC current, Circuit 2 at offshore sub. (A)				164.57	
Output DC voltage at the offshore sub. (kV)				421.93	
Output DC current at the offshore sub. (A)				413.36	
Input DC voltage at the onshore sub. (kV)				420.22	
Input DC current at the onshore sub. (A)				413.36	
Power and losses in the nacelle and tower					
Turbine power (kW)				3600	
HG output power (kW)				3492	
Rectifier output power (kW)				3474.54	
HG efficiency (%) and losses (kW)				97	108
Rectifier efficiency (%) and losses (kW)				99.5	17.46
Tower cable power loss (kW)				0.14	
Turbine link cable					
From–to		1 to 2	2 to 3	3 to 4	4 to 5
Inter-array cable loss (kW)		0.32	1.30	2.92	5.20
Cable loss, Circuit 2 to the offshore substation (kW)				1.49	
Input power and losses at offshore sub. from Circuit 2					
Power (kW)	17,360.76	Losses (kW)		639.24	
Input power and losses at offshore sub. from Circuits 1–6					
Power (kW)	177,079.75	Losses (kW)		6520.25	
DC/DC converters efficiency at offshore sub. (%)				98.5	
Loss at offshore substation (kW)				2656.20	
Cable losses, offshore to onshore sub. (kW)				717.66	
Total power at onshore substation (kW)				173,705.90	
Total losses at onshore substation (kW)				9894.10	
Total system efficiency (%)				94.61	

Power flow results for the DC system at three-quarter, half-, and one-quarter loads are presented in Tables 3.18, 3.19, and 3.20, respectively. Voltage drop at the tower base from Turbine 1 to the offshore substation at different loads is:

$$1.\ \text{Full-load}: \left(\frac{V_1 - V_6}{V_1}\right) 100 = \left(\frac{105.6 - 105.49}{105.6}\right) 100 = 0.1\%. \qquad (3.28)$$

Table 3.17 Rotor speed, HG voltage, and DC/DC converter ratio at different loads

Loading			Rotor speed		HG voltage		Offshore substation voltage kV	DC/DC converter step-up ratio
Item	p.u.	kW	p.u.	RPM	p.u.	kV		
Full	1.00	3600	1.00	13	1.00	105.6	420	4
Three-quarter	0.75	2700	0.91	11.83	0.91	96.1		4.4
Half	0.5	1800	0.79	10.27	0.79	83.42		5
One-quarter	0.25	900	0.63	8.19	0.63	66.53		6.3
Cut-in	0.05	180	0.38	5	0.38	40.13		10.5

Table 3.18 DC system three-quarter-load analysis

Turbine					
Turbine no.	1	2	3	4	5
HG phase voltage (kV)	34.68	34.67	34.67	34.66	34.66
HG DC voltage (kV)	96.10	96.09	96.07	96.05	96.05
HG DC current output (A)	27.12	27.12	27.13	27.13	27.13
DC voltage at the tower base, kV	96.09	96.09	96.07	96.05	96.01
Interconnection and transmission					
Input DC voltage at the offshore substation (kV)	96				
Input DC current, Circuit 2 at offshore sub. (A)	135.63				
Output DC voltage at the offshore sub. (kV)	422.42				
Output DC current at the offshore sub. (A)	309.70				
Input DC voltage at the onshore sub. (kV)	421.12				
Input DC current at the onshore sub. (A)	309.70				
Power and losses in the nacelle and tower					
Turbine power (kW)	2700				
HG output power (kW)	2619				
Rectifier output power (kW)	2605.91				
HG efficiency (%) and losses (kW)	97	81			
Rectifier efficiency (%) and losses (kW)	99.5	13.10			
Tower cable power loss (kW)	0.10				
Turbine link cable					
From–to	1 to 2	2 to 3	3 to 4	4 to 5	
Inter-array cable loss (kW)	0.22	0.88	1.99	3.53	
Cable loss, Circuit 2 to the offshore substation (kW)	1.01				
Input power and losses at offshore sub. from Circuit 2					
Power (kW)	13,021.41	Losses (kW)	478.59		
Input power and losses at offshore sub. from Circuits 1–6					
Power (kW)	132,818.43	Losses (kW)	4881.57		
DC/DC converters efficiency at offshore sub. (%)	98.5				
Loss at offshore substation (kW)	1992.28				
Cable losses, offshore to onshore sub. (kW)	402.85				
Total power at onshore substation (kW)	130,423.31				
Total losses at onshore substation (kW)	7276.69				
Total system efficiency (%)	94.72				

Table 3.19 DC system half-load analysis

Turbine					
Turbine no.	1	2	3	4	5
HG phase voltage (kV)	30.1	30.1	30.1	30.09	30.09
HG DC voltage (kV)	83.43	83.42	83.40	83.39	83.39
HG DC current output (A)	20.82	20.83	20.83	20.83	20.83
DC voltage at the tower base, kV	83.42	83.41	83.40	83.39	83.36
Interconnection and transmission					
Input DC voltage at the offshore substation (kV)				83.35	
Input DC current, Circuit 2 at offshore sub. (A)				104.12	
Output DC voltage at the offshore sub. (kV)				416.77	
Output DC current at the offshore sub. (A)				209.28	
Input DC voltage at the onshore sub. (kV)				415.89	
Input DC current at the onshore sub. (A)				209.28	
Power and losses in the nacelle and tower					
Turbine power (kW)				1800	
HG output power (kW)				1746	
Rectifier output power (kW)				1737.27	
HG efficiency (%) and losses (kW)				97	54
Rectifier efficiency (%) and losses (kW)				99.5	8.73
Tower cable power loss (kW)				0.06	
Turbine link cable					
From–to		1 to 2	2 to 3	3 to 4	4 to 5
Inter-array cable loss (kW)		0.13	0.52	1.17	2.08
Cable loss, Circuit 2 to the offshore substation (kW)				0.6	
Input power and losses at offshore sub. from Circuit 2					
Power (kW)	8681.57	Losses (kW)		318.43	
Input power and losses at offshore sub. from Circuits 1–6					
Power (kW)	88,551.99	Losses (kW)		3248.01	
DC/DC converters efficiency at offshore sub. (%)				98.5	
Loss at offshore substation (kW)				1328.28	
Cable losses, offshore to onshore sub. (kW)				183.96	
Total power at onshore substation (kW)				87,039.75	
Total losses at onshore substation (kW)				4760.25	
Total system efficiency (%)				94.81	

$$2.\ \text{Three-quarter load}: \left(\frac{V_1 - V_6}{V_1}\right) 100 = \left(\frac{96.09 - 96}{96.09}\right) 100 = 0.09\%. \quad (3.29)$$

$$3.\ \text{Half load}: \left(\frac{V_1 - V_6}{V_1}\right) 100 = \left(\frac{83.42 - 83.35}{83.42}\right) 100 = 0.08\% \quad (3.30)$$

Table 3.20 DC system one-quarter-load analysis

Turbine					
Turbine no.	1	2	3	4	5
HG phase voltage (kV)	24.01	24	24	24	24
HG DC voltage (kV)	66.53	66.52	66.52	66.50	66.50
HG DC current output (A)	13.06	13.06	13.06	13.06	13.06
DC voltage at the tower base, kV	66.53	66.52	66.51	66.50	66.49
Interconnection and transmission					
Input DC voltage at the offshore substation (kV)				66.48	
Input DC current, Circuit 2 at offshore sub. (A)				65.30	
Output DC voltage at the offshore sub. (kV)				418.85	
Output DC current at the offshore sub. (A)				104.13	
Input DC voltage at the onshore sub. (kV)				418.42	
Input DC current at the onshore sub. (A)				104.13	
Power and losses in the nacelle and tower					
Turbine power (kW)				900	
HG output power (kW)				873	
Rectifier output power (kW)				868.63	
HG efficiency (%) and losses (kW)				97	27
Rectifier efficiency (%) and losses (kW)				99.5	4.37
Tower cable power loss (kW)				0.02	
Turbine link cable					
From–to		1 to 2	2 to 3	3 to 4	4 to 5
Inter-array cable loss (kW)		0.05	0.02	0.46	0.82
Cable loss, Circuit 2 to the offshore substation (kW)				0.23	
Input power and losses at offshore sub. from Circuit 2					
Power (kW)	4341.30	Losses (kW)		158.70	
Input power and losses at offshore sub. from Circuits 1–6					
Power (kW)	44,281.21	Losses (kW)		1618.79	
DC/DC converters efficiency at offshore sub. (%)				98.5	
Loss at offshore substation (kW)				664.22	
Cable losses, offshore to onshore sub. (kW)				45.54	
Total power at onshore substation (kW)				43,571.45	
Total losses at onshore substation (kW)				2328.55	
Total system efficiency (%)				94.93	

$$4.\ \text{one-quarter load}: \left(\frac{V_1 - V_6}{V_1}\right) 100 = \left(\frac{66.53 - 66.48}{66.53}\right) 100$$

$$= 0.075\%. \tag{3.31}$$

Therefore, as the load decreases, the voltage drop from the turbines to the offshore substation reduces as result of lower currents flowing the system. Note that for all loads, the voltage decreases from Turbine 1 to the offshore substation, i.e.:

Table 3.21 Comparison between Walney and DC system

Item	Walney 1	DC system
Voltage scenario	Interconnections: variable Transmission: fixed	Interconnections: variable Transmission: fixed
Full load		
Total installed capacity (kW)	183,600	183,600
Total power (kW)	167,346.54	173,705.90
Total system loss (kW)	16,477.50	9894.10
Efficiency %	91.15	94.61
Three-quarter load		
Total power (kW)	125,792.79	130,423.31
Total system loss (kW)	11,998.32	7276.69
Efficiency %	91.35	94.72
Half load		
Total power (kW)	83,822.82	87,039.75
Total system loss (kW)	8001.40	4760.25
Efficiency %	91.31	94.81
One-quarter load		
Total power (kW)	41,415.46	43,571.45
Total system loss (kW)	4485.71	2328.55
Efficiency %	90.23	94.93

$$V_1 > V_2 > V_3 > V_4 > V_5 > V_6$$
$$V_7 > V_8 \tag{3.32}$$

Therefore, a power flow from turbines to the grid occurs at all loads.
The loss analysis at full load shows that:

- The total loss for the DC system is 9894.10 kW.
- Circuits 1–6 loss contributes to 65.9% of the total loss.
- The offshore substation DC/DC converter loss contributes to 26.38% of the total system loss.
- The losses in the cable connecting the offshore to onshore substation contribute to 7.25% of the total system loss.

The system total efficiency for different loads is:

$$1.\ \text{Full load}: 94.61\%. \tag{3.33}$$

$$2.\ \text{Three-quarter load}: 94.72\%. \tag{3.34}$$

$$3.\ \text{Half load}: 94.81\%. \tag{3.35}$$

$$4.\ \text{One-quarter load}: 94.93\%. \tag{3.36}$$

Table 3.22 Mass comparison

Item	Walney system		DC system	
	Mass (tonnes)	Mass (%)	Mass (tonnes)	Mass (%)
Power train, turbine, and tower mass				
Blades × 3	54	12.13	54	12.42
Hub mass	38.5	8.65	38.5	8.85
Shaft	27.5	6.18	27.5	6.32
Gearbox	42.5	9.55	28.3	6.5
IG	10	2.25	–	–
HG	–	–	24.2	5.6
Two VSCs	2.4	0.54	–	–
Rectifier	–	–	0.25	0.06
Turbine transformer	8	1.80	–	–
Blades and hub bolts	2.16	0.49	2.16	0.50
Tower[a]	260	58.42	260	59.78
Total	445.06	100	434.91	100
Cable mass[b]				
Tower cable	4.78	0.49	0.03	0.01
Inter-array cables between turbines	133.06	13.56	12	5.06
Cable, circuits to offshore	91.24	9.3	9.89	4.17
Cable, offshore sub. to shore	751.89	76.65	215.33	90.76
Total	980.97	100	237.25	100

[a]83.5 m above mean sea level
[b]Copper density is taken as 8960 kg/m^3

The system efficiency is similar at different loads. Hence, similar benefits and system performance are seen at all loads. Note that the HG, rectifier, and DC/DC converter efficiencies are constant at all loads.

3.6 Comparison Between DC System and Walney Wind Farm

Table 3.21 compares results of the DC system and Walney wind farm in terms of total power, losses, and efficiency at different loads. The DC system increases the wind farm efficiency by 3.8% at full load. Also, the system efficiency is increased by 3.7%, 3.8%, and 5.2% at three-quarter, half-, and one-quarter loads, respectively. Therefore, the DC system results in similar benefits at all loads compared to the Walney system. Note that the voltage throughout the system for the Walney is fixed. However, the interconnection voltage varies in the DC system while the offshore to onshore transmission voltage is fixed. Table 3.22 presents a mass comparison of the DC and Walney systems.

Cable volume is calculated as follows:

$$V_{\text{cable}} = A_{\text{cable}}\ l_{\text{cable}}\left(\text{m}^3\right) \tag{3.37}$$

The cable mass is then calculated as:

$$\text{Mass}_{\text{cable}} = \delta_{\text{cu}}\ V_{\text{cable}}(\text{kg}) \tag{3.38}$$

where

A_{cable} is cable cross section (m^2).
l_{cable} is cable length (m).
δ_{cu} is the copper density taken as 8960 (kg/m^3).

Mass for different cables in the Walney and DC system is listed in Table 3.22. The analysis demonstrates the improvements of the DC system when compared to the benchmark Walney 1. For example, the DC system results in 16% and 2.3% mass reduction for the power train and total turbine structure, respectively. The power train in Walney includes a gearbox, an IG, two VSCs, and a transformer, i.e., five series connected components, while for the DC system, it includes a gearbox, an HG, and a rectifier, i.e., three series connected components. Therefore, the DC system results in a lower component count. Note that the DC system uses DC/DC converters at the offshore platform as compared with two transformers. In addition, the cable system copper mass is 75.8% lower.

3.7 Summary

A DC wind generation scheme based on a high-voltage HG, HVDC interconnection, and transmission is discussed in this chapter. The DC system is compared with a commercial system, Walney 1, chosen as a representative benchmark of existing industry practice. Analysis results from the two systems show that the DC system results in 3.8% higher energy conversion efficiency assuming all turbines are operating at, or near, their full-load thermal rating. Results for the DC wind generation scheme at other loads have also been assessed and show similar efficiency gains and benefits.

The presented system highlights the potential reduction of system plant within the nacelle and turbine tower, features that will improve the serviceability and maintenance of the wind farm over time. The DC scheme has 2.3% lower mass for the turbine and 75.8% lower mass for the cable systems. These are achieved by using higher voltage for generation, interconnection, and transmission.

However, the major contribution of the DC system is the elimination of active VSCs within the nacelle and tower and replacing them with passive rectifiers built into the HG or located close to the machine terminals. The HG uses two rotor field excitations, PM and WF, with a nine-phase stator that yields improved power quality while essentially eliminating the electrolytic smoothing devices (DC-link capacitance) that would otherwise be required.

The VSCs and electrolytic capacitors are major causes of variable speed drive system failures, hence the claim for improved serviceability, maintenance, and lower operational cost. It is discussed that the turbine interconnection voltage is allowed to vary (but controlled), compared to the essentially fixed interconnection voltage for the benchmark AC system. For the DC scheme, the active power electronic converters (high-voltage, high-power DC/DC converters) are located at the offshore substation where access for maintenance and repair is easier and more cost-effective. It is envisaged that these savings will make a significant contribution to overall cost reduction of the wind farm and system, although this is not assessed nor quantified here.

Chapter 4
Hybrid Generator (HG) Concept

4.1 Electric Machines and Converters for Wind Turbines

4.1.1 Doubly Fed Induction Generators (DFIGs)

Different turbine generator topologies, both direct drive and geared, have been proposed in the literature and developed by manufactures for wind turbine applications. Figure 4.1 shows a doubly fed induction generator (DFIG) connected to the grid via two back-to-back VSCs that can be controlled to operate in four quadrants.

The DFIG has a three-phase stator and wound rotor windings. The four-quadrant VSCs connect the rotor variable voltage and frequency output back to the fixed voltage and frequency main supply via a coupling transformer and enable the control of both the generator torque and reactive power flow. The rotor speed is controlled by absorbing or injecting active power to the DFIG rotor. A typical DFIG speed range is $\pm 30\%$ around synchronous speed, and since only a portion of the generated active power flows through the rotor, the DFIG-side VSC power ratings are only around 30% of the generator rating. However, usually, the VSCs are oversized due to increased demand in reactive power controllability, wider speed range, and low-voltage ride through requirements.

4.1.2 VSC-Coupled Induction, Synchronous, and PM Generators

In order to achieve a wider operating range of speed, the induction generator (IG), permanent magnet (PM), or wound field synchronous generator (SG) are connected to the grid via two fully rated converters in a back-to-back configuration. Figure 4.2 shows a squirrel cage rotor IG in a wind turbine connected to the grid via two back-to-back VSCs. Note that the rating of each converter in this configuration has to be

© Springer Nature Switzerland AG 2020

O. Beik, A. S. Al-Adsani, *DC Wind Generation Systems*,
https://doi.org/10.1007/978-3-030-39346-5_4

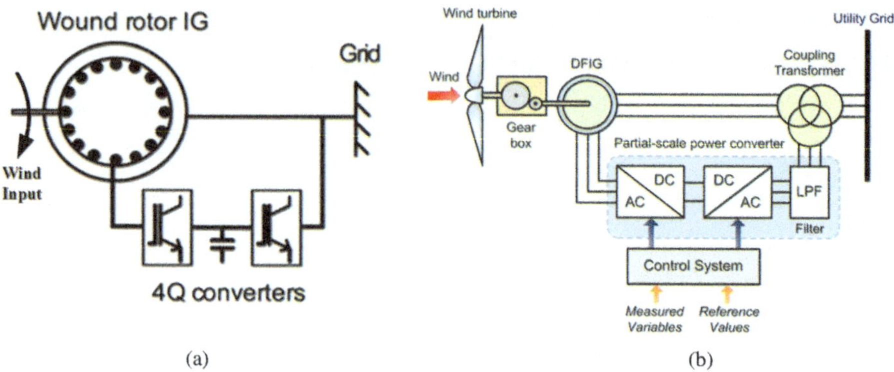

Fig. 4.1 DFIG configuration (**a**) Simplified schematic (replace with a drawing) (**b**) Geared DFIG

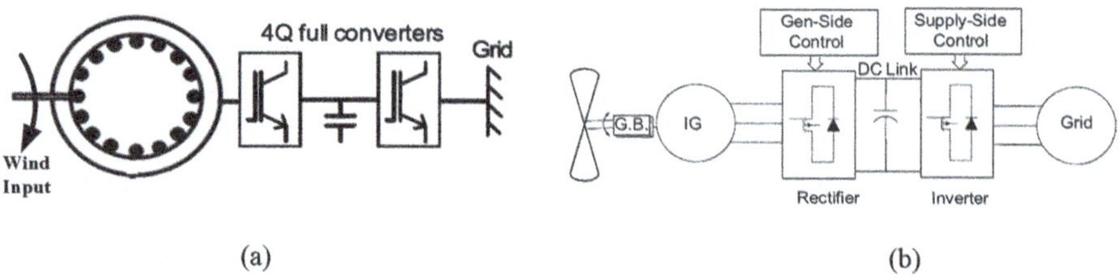

Fig. 4.2 IG connected to grid via two fully rated VSC (**a**) Simplified schematic (to be drawn) (**b**) Geared IG

the rating of the generator since the total generator power flows through each converter.

The IG is a mature technology and is relatively inexpensive and robust and requires low maintenance. In the configuration shown in Fig. 4.2, the IG-side converter controls the IG speed via vector control to extract maximum wind power. The grid-side converter performs active and reactive power control. A comparison in terms of hardware components, operation, and energy output between DFIG and IG is presented in [58]. There are pros and cons for each scheme; for example, the energy output from the DFIG is limited due to limited speed variation range. However, the converters in DFIG schemes are not fully rated, so it is a lower-cost scheme. This said, most large wind farms are employing the dual converter scheme of Fig. 4.2, as per the Walney farms due to the increased energy conversion capability. For such large wind farms as the Walney, lifetime energy conversion clearly drives system choice as opposed to initial equipment capital costs. Figure 4.3 presents a wind conversion scheme employing a PM generator connected to the grid via two fully rated VSCs. The functionality of the conversion system in Fig. 4.3 is similar to that of Fig. 4.2; hence, the machine-side converter controls the rotor speed to extract a maximum power from wind, while the grid-side converter controls the active and reactive power flow and maintain a fixed DC-link voltage.

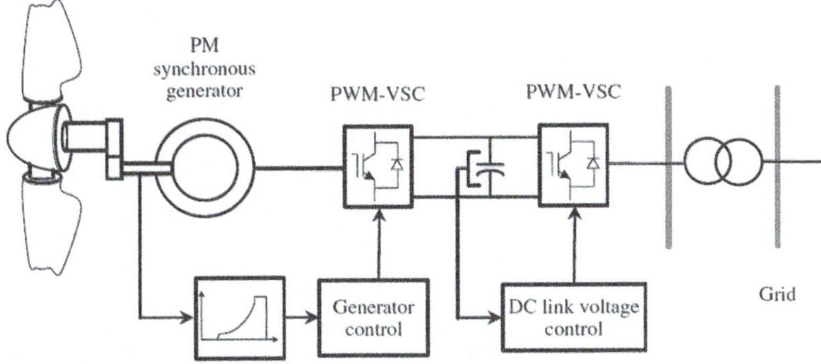

Fig. 4.3 PM generator connected to grid via two fully rated VSC

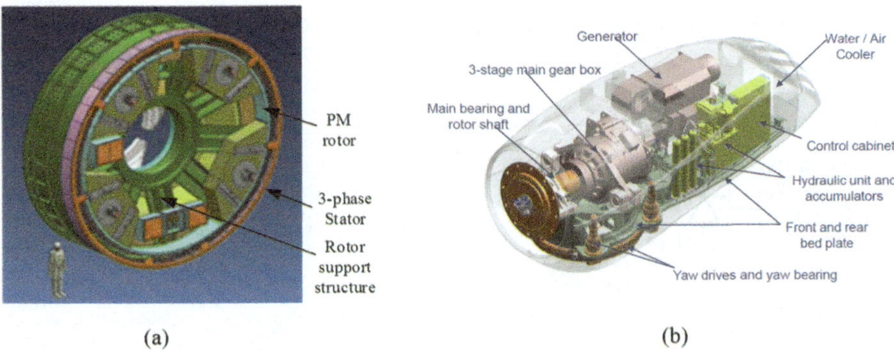

Fig. 4.4 Direct drive PM and geared DFIG (**a**) Alstom 6 MW direct drive PM generator [59]. (**b**) 2 MW geared DFIG from DeWind D8.2 wind turbine [45]

PM machines have higher flux-densities compared to IGs and DFIGs; they are usually more power dense compared to other generator technologies, do not need external supply and slip rings, and avoid reactive power compensation. Figure 4.4 shows a 6 MW direct drive PM generator for the Alstom Haliade 150-6 MW wind turbine [59], where the machine has larger diameter compared to its axial length due to low rotor speed, that is, maximum 11.5 RPM. However, the 2 MW geared DFIG from DeWind D8.2 wind turbine [45] with a nominal rotor speed of 1700 RPM has a longer axial length compared to its diameter. Figure 4.5 shows another scheme where the PM generator is connected to the grid via a passive rectifier, a DC/DC converter (chopper/booster), and a VSC. In this scheme, the PM generator vector control is implemented thought the VSC, while the DC/DC converter maintains a fixed DC-link voltage. Note that current industrial turbines are controlled to operate in the limited speed operation regime (as discussed in Chap. 2); hence, the PM generator maximum speed is the machine rated speed, and no field weakening of the machine is required.

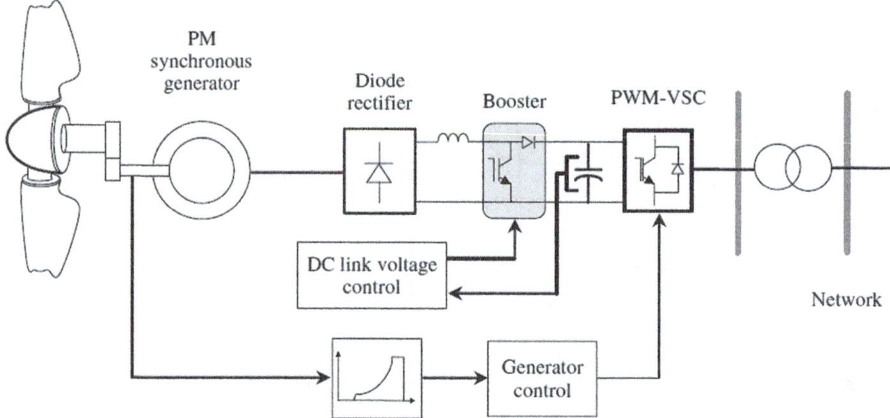

Fig. 4.5 PM generator connected to grid via passive rectifier, DC/DC converter, and a VSC

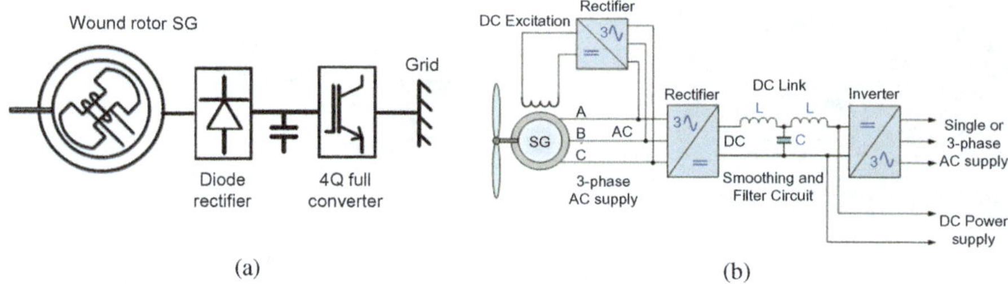

Fig. 4.6 Wound field SG connected to grid via passive rectifier and a fully rated VSC (**a**) Simplified model. (Draw it) (**b**) Direct drive SG

Figure 4.6 shows a wound field SG connected to the grid via a passive rectifier and a VSC. Using the wound field SG, the machine-side VSC that was required for PM generator is eliminated due to capability of the SG to have an active output voltage control via controlling the machine rotor field. However, in order to improve and facilitate the system control, a DC/DC converter can be used as in the system shown in Fig. 4.5. The wound field SG is capable of providing its own field excitation by means of a brushless exciter mounted on its shaft. Unlike the PM generator, the wound field SG output voltage can be controlled via controlling the machine field current independent of the load. However, for the same output power, wound field SGs are normally heavier than PM rotors, are typically bulkier, and have higher rotor losses.

4.1.3 Comparison of Generators

A theoretical comparison of different turbine generator topologies is presented by Polinder et al. in [43]. Polinder studies five different generator systems for a 3 MW rated output power and a 15 RPM turbine shaft speed. The generators include:

- A DFIG with a three-stage gearbox (DFIG3G) and gear ratio of 1:80, that is, a machine speed of 1200 RPM
- A DFIG with a single-stage gearbox (DFIG1G) and a gear ratio of 1:6, that is, a machine speed of 90 RPM
- A direct drive wound field SG (DDSG) with nominal speed of 15 RPM
- A direct drive PM generator (DDPMG) with nominal speed of 15 RPM
- And a PM generator with single-stage gearbox (PMG1G) and gear ratio of 1:6, that is, a machine speed of 90 RPM

The comparison is based on machine mass, energy yield, losses, and cost. Table 4.1 summarizes Polinder's comparison. Note that these comparisons cannot be generalized. For example, the DFIG3G is the lightest, but for a fair mass comparison, the mass of gearbox–generator–converter systems needs to be taken into account since the direct drive machine is normally heavier. Moreover, technical characteristics and benefits need to be investigated for a complete analysis. Data for a HV generator used as a benchmark validation is also included in Table 4.1 to confirm

Table 4.1 Comparison of different generators for wind turbine application as reported in [43] and the benchmark HV generator

Item	DFIG3G	DDSG	DDPMG	PMG1G	DFIG1G	Benchmark SG
Stator radius (m)	0.42	2.5	2.5	1.8	1.8	2.7
Active axial length (m)	0.75	1.2	1.2	0.4	0.6	0.4
Air-gap thickness (mm)	1	5	5	3.6	2	7
Mass[a] (tons)	5.25	45.1	24.1	6.11	11.37	7.02
Cost[b] (kEuros)	1870	2117	1982	1883	1837	–
Annual energy yield (GWh)	7.73	7.88	8.04	7.84	7.80	–
Turbine speed (RPM)	15					–
Rated generator power (MW)	3					5.52
Machine speed (RPM)	1200	15	15	90	90	600
Gearbox ratio	1:80	1:1	1:1	1:6	1:6	–
Gearbox mass for the Siemens SWT-3.6-107 wind turbine gear ratio (1:119)						
42.5 (tons)						

[a]Mass includes mass of iron, copper, and PM material only for the benchmark SG mass does not include mass of rotor iron

[b]Cost includes cost of generator material, construction, gearbox, power electronics, and other wind turbine parts

machine suitability. Gearbox mass for the Siemens SWT-3.6-107 wind turbine that is used in Walney wind farm is also stated in Table 4.1.

4.2 Hybrid Generator (HG) Concept

Section 3.5 discussed a DC offshore wind generation system that could potentially result in reduced system cost due to reduced component count and complexity. The DC system uses a multiphase HG, the AC output of which is passively rectified for transmission from the turbine tower. The HG is an AC machine having a multiphase, high-voltage winding and uses two rotor sections, namely, a permanent magnet (PM) rotor and a wound field (WF) rotor, to provide the total machine excitation. Hence, the name hybrid generator refers to the combination of these two rotors excitation schemes. The PM and WF rotor sections exist on one rotor assembly inside one machine housing. Thus, the rotors rotate with the same speed. A schematic view of the HG is depicted in Fig. 4.7a. DC current for the WF rotor excitation is provided via a brushless exciter system common to industrial synchronous machine systems. Therefore, the HG combines the output voltage due to a fixed field from the PM rotor and a controlled variable voltage due to the variable field of the WF rotor. The total machine output voltage is the sum of the voltages due to PM and WF sections. The WF therefore contributes a controllable variable voltage at the stator output over a limited range. The range of the total voltage variation depends on the design of the WF rotor of the HG. Figure 4.7b shows the HG in a wind turbine system.

In general, PM machines have a fixed rotor field excitation that induces a fixed voltage into the stator winding at a fixed speed. Hence, in order to control the machine output voltage/current, a fully rated power electronic converter is required. However, if the power electronic converter is eliminated, an alternative solution to regulate machine output voltage and hence power is the HG configuration with a passive rectifier, shown schematically in Fig. 4.7a, b for a wind turbine application. Referring to Fig. 4.7b, the output voltage of the machine induced by the PM

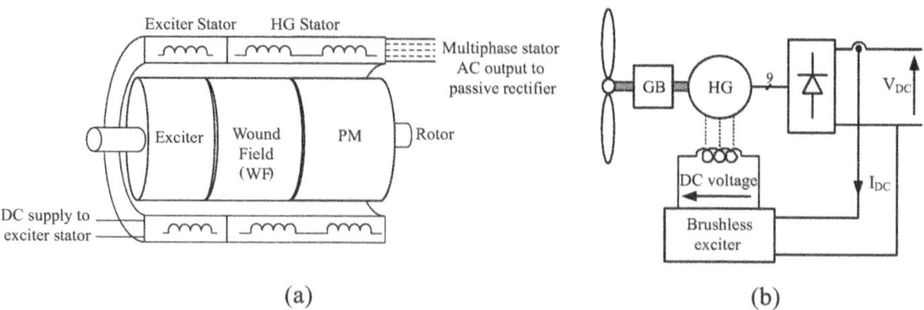

Fig. 4.7 Hybrid generator (HG) concept (**a**) HG schematic view (**b**) HG in a wind turbine

excitation is regulated by the WF excitation. The WF is fed a DC current through a brushless exciter and a rotating rectifier. The brushless exciter has a DC stator that is fed through the output of the HG as shown in Fig. 4.7b. The machine terminal voltage is converted to a DC voltage via a passive rectification stage. The brushless exciter has an AC rotor that may use a three- or multiphase configuration. Output of the brushless exciter rotor is then connected to a rotating rectifier that in turn provides the DC supply to the HG WF. A multiphase design for the brushless exciter can give good DC output quality with lower voltage ripple and higher output voltage than compared to three-phase exciter, hence eliminating the need for passive capacitors at its output.

4.3   High-Voltage, Three-Phase Benchmark Synchronous Generator (SG)

4.3.1   General Details

The HG is a high-voltage, high-power machine. Due to unusual and unique structure of the HG and in order to have a sensible and viable design consistent with industry practice, a standard high-voltage SG is designed within the normal design practice and constraints and set as a benchmark machine. The design of the HG is then developed based on the benchmark SG specifications.

The electromagnetic characteristics of the benchmark SG are investigated using finite element analysis (FEA) to obtain machine electromagnetic field distribution and flux-linkages. Analyses are carried out for no-load and full-load conditions where the back-EMF, torque, and machine performance are simulated. Once the benchmark SG characteristics are obtained, they are used as acceptable measures for the subsequent HG design discussed in Chap. 5.

The benchmark SG is a three-phase, 11 kV (line-to-line RMS voltage), synchronous generator designed for a hydro generation plant. The benchmark SG is a salient pole machine rated at 5.52 MW with a nominal speed of 600 rpm. The machine is directly connected to the three-phase, 50 Hz grid. Table 4.2 lists the main machine specifications. The high-pole, low-speed design was chosen to be compatible with the HG concept.

The benchmark SG stator outer diameter is 2700 mm with an inner diameter of 2150 mm and active axial length of 400 mm, as illustrated in Fig. 4.8. The benchmark SG main dimensions are summarized in Table 4.3, and for the purpose of comparison, specification details for the IG used in the Walney wind turbines are listed in Table 4.4.

The front-end cross section of the benchmark SG is illustrated in Fig. 4.8 with annotated main dimensions. The stator tooth/slot and rotor pole geometries are presented in Figs. 4.9 and 4.10, respectively. As seen from Fig. 4.10a, the rotor has a salient pole whose "shoe" length varies from 70 mm at the middle of the pole to

Table 4.2 Benchmark SG specifications

Parameter	Description/value
Generator type	Salient pole synchronous generator (SG)
Rated power (MW)	5.52
Rated RMS line-to-line voltage (kV)	11
Rated RMS phase current (A)	362
Field DC current at rated power (A)	486.23
Rated speed (RPM)	600
Number of poles	10
Frequency (Hz)	50
Number of phases	3

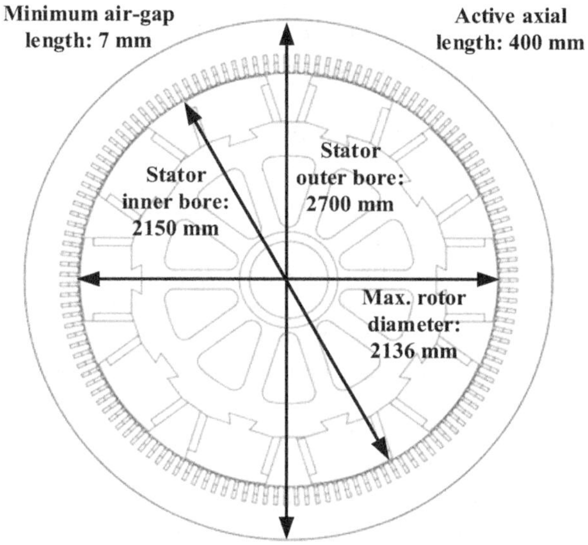

Fig. 4.8 Front-end cross section of the benchmark SG

43.7 mm at the corners. Therefore, the air-gap length changes from a minimum of 7 mm in the middle of the pole to a maximum of 11 mm at the pole corners. Hence, the benchmark SG does not have a constant air-gap over the pole-face, referred to here as "adjusted" air-gap. To test contribution of this design feature, a modified rotor topology was modeled having a constant air-gap at the rotor pole-to-stator interface as illustrated in Fig. 4.10b. It will be shown later that the adjusted air-gap rotor topology reduces cogging torque. The stator slot length is 95 mm out of which 5 mm is the wedge length to keep the winding inside the slot, 3 mm the stator tip, and 87 mm the stator main body. The stator slot width is 18.5 mm and the tooth width is 31.53 mm. Therefore, the stator slot/tooth pitch is 50.03 mm.

Table 4.3 Benchmark SG dimensions

Parameter	Value
Machine axial length (mm)	400
Machine outer diameter (mm)	2700
Machine inner diameter (mm)	2150
Minimum air-gap length (mm)	7
Maximum air-gap length (mm)	11
Maximum rotor diameter of body (mm)	2136
Shaft diameter (mm)	400
Bearing diameter (mm)	250
Bearing length (mm)	200
Rotor pole body height (mm)	195 mm
Rotor pole body width (mm)	350 mm
Rotor pole shoe minimum height (mm)	43.7 mm
Number of stator slots/teeth	135
Stator tooth tip width (mm)	31.35
Stator slot width (mm)	18.50
Stator slot length from air-gap (mm)	90
Stator slot length for winding (mm)	87
Wedge length (mm)	5
Stator back-iron thinness (mm)	180
Stator slot area available for winding (mm^2)	1609.5

Table 4.4 Walney wind farm IG specifications

Parameter	Description
Generator type	IG
Rated power (MW)	3.6 MW
Rated voltage (V)	690
Rated current (A)	3075.7
Rated speed (RPM)	1500
Number of poles	4
Frequency (Hz)	50
Number of phases	3
Machine length (m)	2.17
Machine outer diameter (m)	1.15
Machine inner diameter (m)	0.57

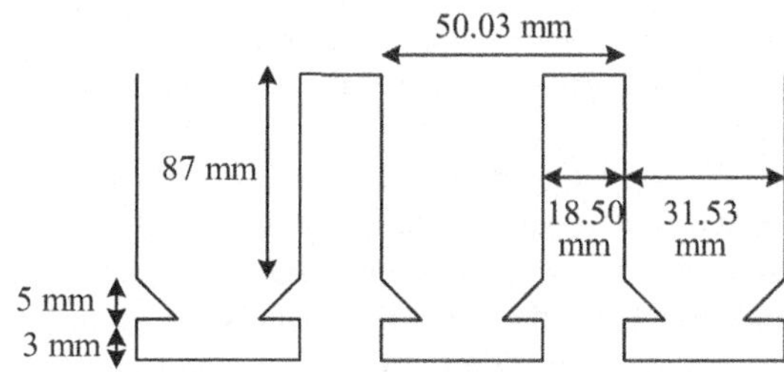

Fig. 4.9 Benchmark SG stator tooth/slot details

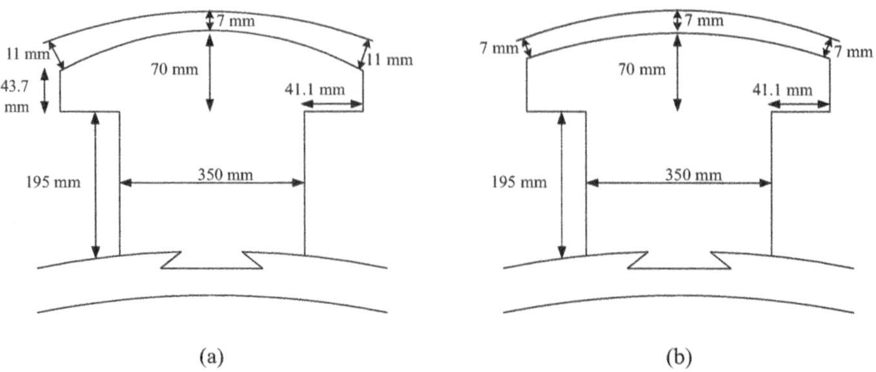

Fig. 4.10 Benchmark SG rotor pole structure (**a**) Benchmark machine rotor with "adjusted" air-gap (**b**) Constant air-gap rotor topology to test "adjusted" air-gap feature

4.3.2 Winding Arrangement

The benchmark SG stator has 135 slots with a three-phase distributed winding. The pole pitch (number of slots per pole) is 13.5 slots, while the coil pitch (span) is 12 slots. Therefore, since the coil pitch is less than the pole pitch, the winding is short pitched. The winding phase band (sometimes called phase belt) is:

$$\text{Phase band} = \text{slots/pole/phase} = 135/10/3 = 4.5 \tag{4.1}$$

Since the phase band is not an integer, the winding is a double-layer fractional-slot winding. In order to form a fractional-slot winding, a pattern needs to be found. Over the years, motor designers have developed a set of rules that will produce a balanced three-phase fractional-slot winding [tables can be provided if necessary]. Therefore, a pattern of five and four coils for the benchmark SG stator winding is found as illustrated in Fig. 4.11. A set of five coils are connected in series where one side of the coil is placed into the top layer and the other side is in the bottom layer. The four-set coils are also laid out similarly, and the two sets, five-set and four-set, are then connected in series to form a complete pattern. Figure 4.11 presents the winding pattern for one phase, and this pattern being repeated five times for all the 135 slots to form the completer phase winding. Therefore, there are 45 coils per phase which are connected in series. Figure 4.12 shows connection of the coils in one phase. The pattern is repeated for the other two phases in the final three-phase winding.

In each stator slot, there are two coils. Each coil consists of four conductors (turns), and each conductor has a total of six strands electrically connected in parallel. The cross section of a stator slot including cross section of coils, conductors, and strands for the benchmark SG is shown in Fig. 4.13. The two coils in the slot are separated using a 4 mm insulating material. The insulation systems for the windings will be discussed in Chap. 6.

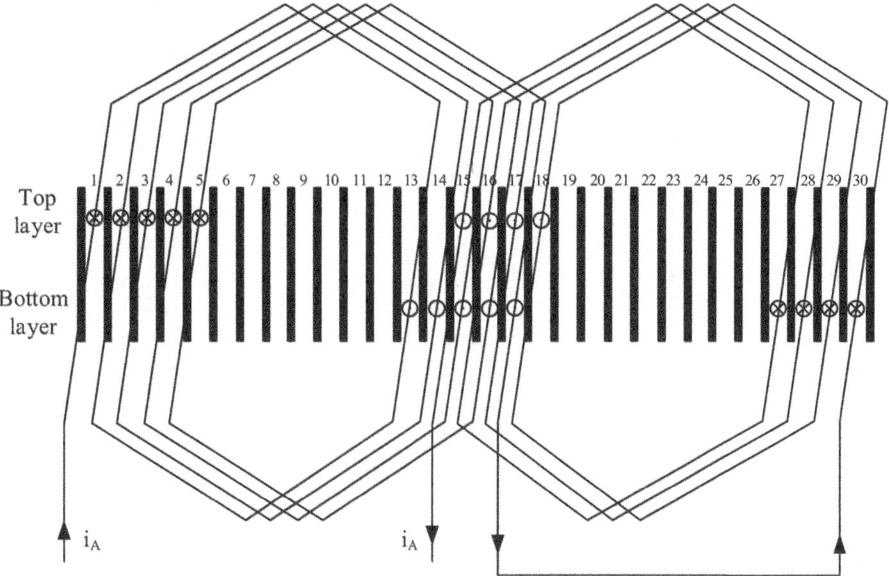

Fig. 4.11 Fractional-slot, short-pitched stator winding for the benchmark SG

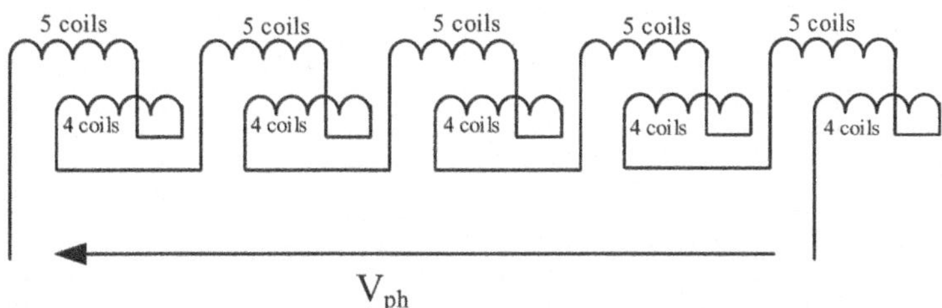

Fig. 4.12 Stator phase coil connections for benchmark SG

Fig. 4.13 Stator slot
showing cross section of
coil, conductor, and strands

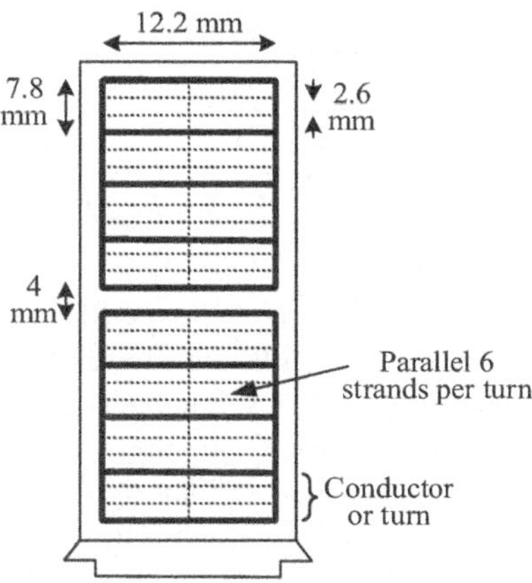

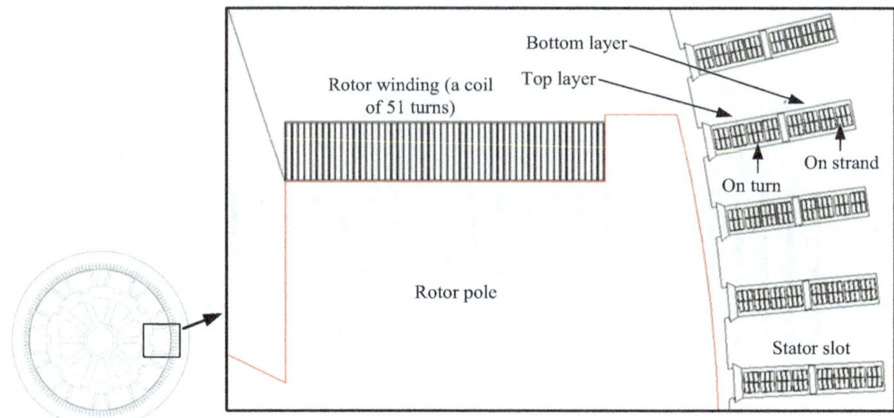

Fig. 4.14 Rotor and stator windings

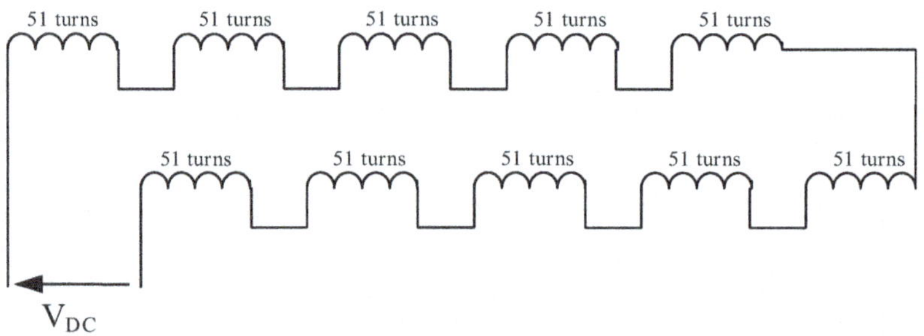

Fig. 4.15 Rotor coil connections for benchmark SG

The DC benchmark SG rotor winding consists of 10 coils, each wound around one salient pole. Each coil has 51 turns and the 10 coils are connected in series. Figure 4.14 shows the rotor winding as well as stator slot windings. The rotor winding connections are shown in Fig. 4.15.

4.3.3 Magnetic Field Distribution and Flux-Density

Using FEA, the benchmark SG open-circuit (O.C.) flux distribution at zero rotor angle and for a full-field current of 486.23 A is shown in Fig. 4.16. It is seen that some parts of the stator back-iron and rotor laminations operate at flux-densities close to 2 T and hence are saturated. Indeed, the machine operates at high flux-density at full load facilitating reduced inductance and hence better power capability. Higher flux-density implies higher iron losses; however, loss management or cooling is accommodated to maintain machine performance.

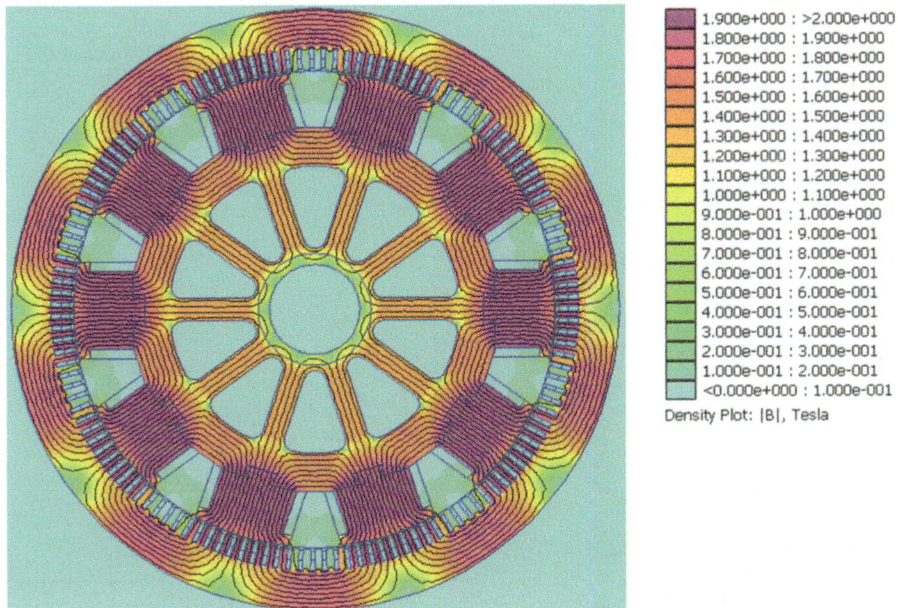

Fig. 4.16 Benchmark SG field distribution at full field

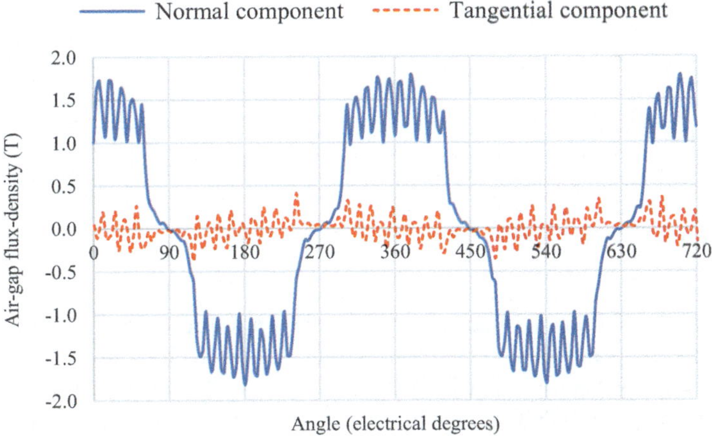

Fig. 4.17 Air-gap flux-density versus rotor electrical angle for two electrical cycles

Figure 4.17 illustrates the normal and tangential component of the air-gap flux-density with respect to rotor electrical angle and for two electrical cycles. The perturbations at the peak flux-density plot are caused by the stator slots and are called "slot effects." The flux crossing the air-gap faces a magnetically uneven surface of slots and teeth combination with variant reluctance which causes the slot effect. Flux-densities for different parts of the machine laminations are listed in

Table 4.5 Flux-density for different machine sections

Section	Flux-density (T)
Average air-gap over half a cycle (normal component)	0.998
Average air-gap over half a cycle (tangential component)	0.11
A	1.69
B	2.1
C	2
D	1.8
E	1.5
F	1.85
G	1.87
H	1.76
I	1.3

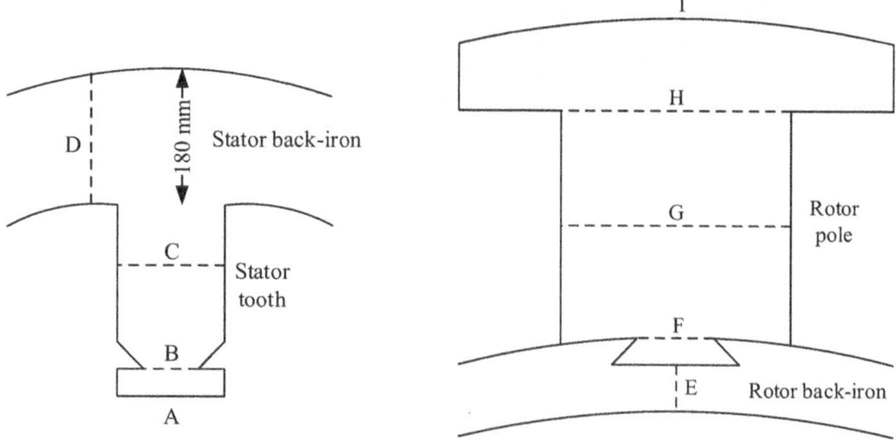

Fig. 4.18 Stator and rotor sections for flux-density measurement

Table 4.5 with reference to machine sections shown in Fig. 4.18. The flux-densities are measured perpendicular to the dashed/solid line for each section.

The average air-gap flux-density over half a cycle that is an indication of machine magnetic loading is 0.998 T (for the normal component). As seen from Table 4.5, the tooth tip operates at the highest flux-density in the machine, that is, 2.1 T. However, the mid-tooth and back-iron flux-densities are 2 T and 1.8 T, respectively. These numbers confirm that the machine operates at high flux-densities at full-field current excitation. In order to define how much the machine is operating into saturation, the FEA analysis at different field currents is performed, and plots for average air-gap, tooth, and back-iron flux-densities are presented in Fig. 4.19 with their value listed in Table 4.6. The tooth flux-density is measured at the middle of the tooth, that is, section "C" in Fig. 4.18. Also, the back-iron flux-density is measured at points with highest saturation.

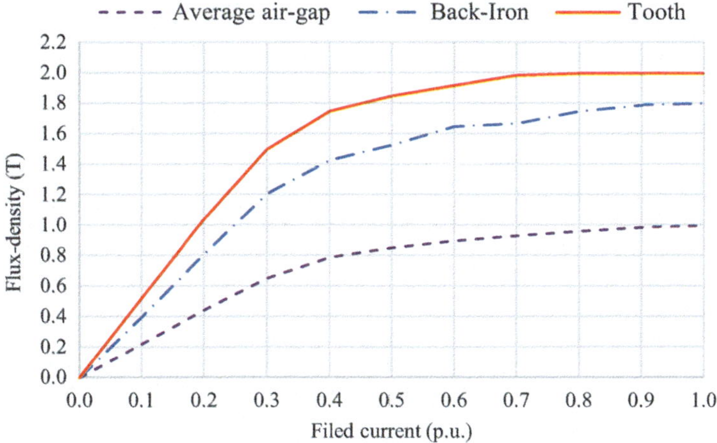

Fig. 4.19 Flux-density versus rotor field current for air-gap, back-iron, and tooth

Table 4.6 Flux-density for different field currents

Field current		Flux-density (T)		
(A)	(p.u.)	Average air-gap	Back-iron	Tooth
0	0.00	0	0	0
48.623	0.10	0.222	0.4	0.52
97.246	0.20	0.445	0.81	1.04
145.869	0.30	0.653	1.21	1.5
194.492	0.40	0.79	1.43	1.75
243.115	0.50	0.854	1.53	1.85
291.738	0.60	0.9	1.65	1.92
340.361	0.70	0.932	1.67	1.986
388.98	0.80	0.962	1.75	2
437.61	0.90	0.988	1.79	2
486.23	1.00	0.998	1.8	2

4.3.4   Flux-Linkage and Back-EMF

The open-circuit flux-linkage over one electrical cycle for one-phase winding and at a full-field current of 486.23 A is shown in Fig. 4.20. As discussed previously, the benchmark SG has an adjusted air-gap, Fig. 4.10a. The flux-linkage from the adjusted air-gap machine is compared with that for the constant air-gap machine shown in Fig. 4.10b, and the results presented in Fig. 4.20, where it can be noted that there is little difference.

The stator phase winding-induced open-circuit back-EMF is calculated from Faraday and Lenz's laws, namely:

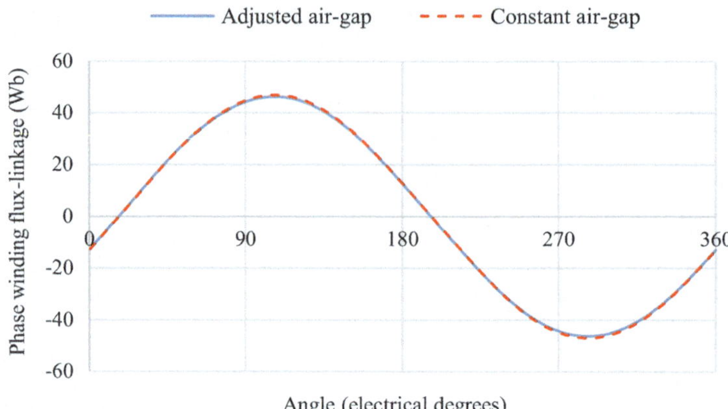

Fig. 4.20 Phase winding open-circuit flux-linkage for the benchmark SG at full field

$$E = -N \frac{d\phi}{dt} \ (\mathrm{V}) \tag{4.2}$$

where ϕ is the flux-linkage per-phase winding. The rate of change in flux-linkage with respect to rotor electrical angle can be expressed in terms of machine speed:

$$\frac{d\phi}{dt} = \frac{d\phi}{d\theta_e} \frac{d\theta_e}{dt}$$
$$\frac{d\theta_e}{dt} = \omega_e \tag{4.3}$$

Therefore, the phase winding open-circuit back-EMF is:

$$E = -N \, \omega_e \, \frac{d\phi}{d\theta_e} \ (\mathrm{V}) \tag{4.4}$$

also,

$$\omega_e = \frac{P}{2} \, \omega_m \ (\text{electrical rad/s}) \tag{4.5}$$

$$\omega_m = \frac{2 \, \pi \, n}{60} \ (\text{mechanical rad/s}) \tag{4.6}$$

where

N: The total number of phase turns
ω_e: Machine electrical rotor speed (rad/s)
ω_m: Machine mechanical rotor speed (rad/s)
θ_e: Rotor electrical angle (rad)
n: Rotor speed (RPM)

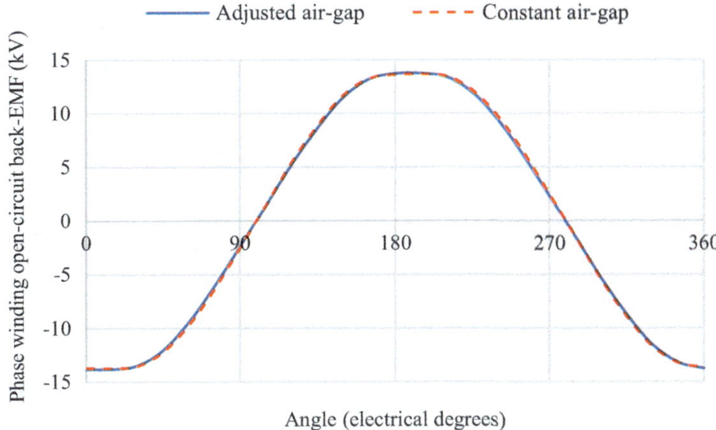

Fig. 4.21 Phase winding open-circuit back-EMF at full-field and nominal rotor speed of 600 RPM

Table 4.7 Phase winding open-circuit back-EMF for different rotor field currents

Field current	Phase back-EMF (kV)	
(A)	Peak	RMS
0.00	0.00	0
30.00	2.03	1.44
60.00	4.09	2.91
90.00	6.14	4.36
122.32	8.25	5.87
149.09	9.80	7.02
193.97	11.32	8.26
235.22	11.98	8.81
298.43	12.66	9.35
382.20	13.29	9.87
486.23	13.77	10.32

P: Number of poles

Figure 4.21 shows the phase winding back-EMF for full-field current and nominal rotor speed of 600 RPM. Here, the back-EMF is plotted for both the adjusted and constant air-gap rotor designs. It is observed that there is no substantial difference in the open-circuit back-EMF between the two topologies. However, it will be shown later that the cogging torque for the adjusted air-gap is significantly smaller. The phase winding open-circuit back-EMF for different rotor currents is shown in Table 4.7, while Fig. 4.22 illustrates these results graphically. Figure 4.22 also confirms that the machine is magnetically saturated at full field. In fact, the linear region only extends to field currents up to 150 A, field current above 150 A progressively moving into the saturated region. Note that the rotor speed is kept at 600 RPM at all field currents. The phase winding open-circuit back-EMF is not a pure sinusoidal waveform, and hence the RMS values are calculated accordingly.

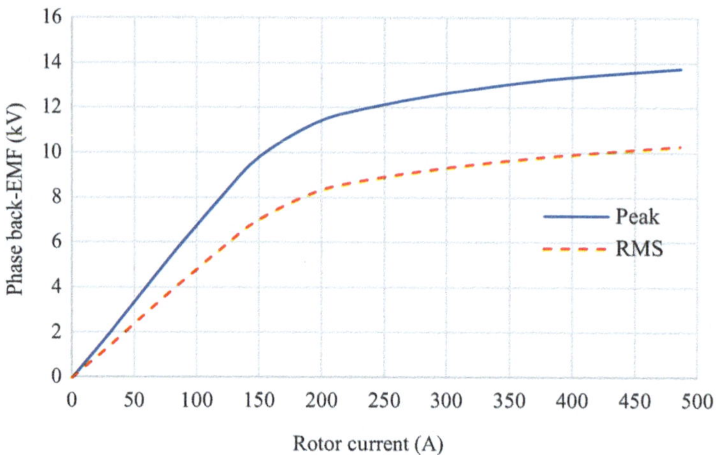

Fig. 4.22 Phase winding open-circuit back-EMF versus rotor field current

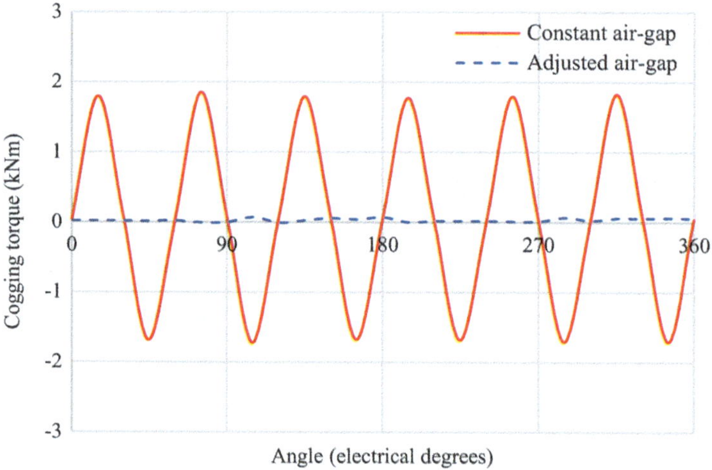

Fig. 4.23 Benchmark SG cogging torque for the adjusted and constant air-gap rotor topologies

4.3.5   Torque Calculations

Cogging torque for the benchmark SG is plotted in Fig. 4.23. The cogging torque is calculated at no-load (zero stator current) and full-field rotor current for both the adjusted and constant air-gap rotor topologies. It is seen that for the adjusted air-gap rotor, the machine cogging torque is substantially lower. Therefore, the effect of adjusting the air-gap is a lower cogging torque in the benchmark SG.

The machine torque versus rotor angle at full-load stator phase winding current (512 A peak) and full-field field current (486.23 A) is presented in Fig. 4.24. The

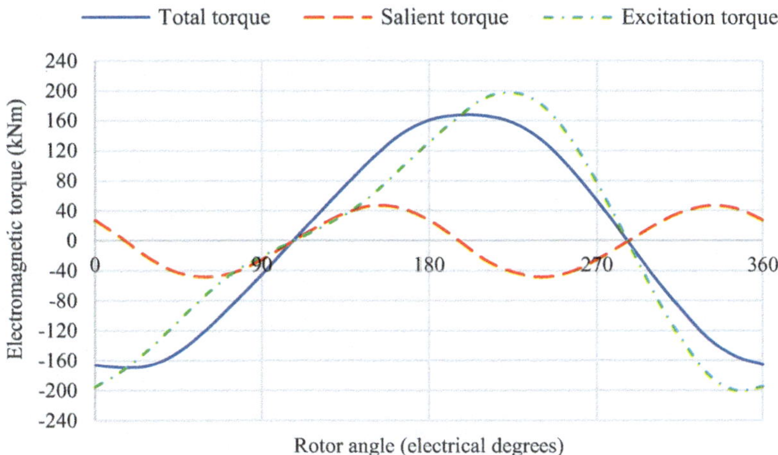

Fig. 4.24 Benchmark SG torque components at full rotor field current and full-load stator winding currents

benchmark SG has a salient rotor; therefore, the total torque has two components: (i) excitation torque and (ii) salient torque. The excitation torque is as a result of the interaction of the rotor and stator magnetic fields. The rotor magnetic field is produced by the rotor field current, and stator magnetic field is produced by the stator three-phase winding currents. The salient torque, however, is only dependent on the machine geometry and the stator currents. As an estimate of each contribution, the excitation torque can be obtained by subtracting the salient torque from the total torque, the results of which is shown in Fig. 4.24. However, the results of Fig. 4.24 are not as would be expected from contemporary machines theory. This feature was investigated further by varying the stator current amplitude and then the rotor field current amplitude. Figure 4.25 presents machine total torque at full field and for full, three-quarter, half, and one-quarter stator current excitation. It is seen that there appears little to no saliency torque component at full rotor field current at any stator loading. In order to see this clearer, Fig. 4.26 shows the near linear variation of peak torque with respect to stator load current. Figure 4.27 shows total torque for different rotor field currents and at full-load stator current excitation. It is seen that as the rotor field current reduces, the effect of saliency, i.e., the twice fundamental component, becomes clear in the total torque. Therefore, as the rotor field current decreases, the machine magnetics move from saturation and the salient torque becomes more apparent in the machine torque versus angle characteristics. Figure 4.28 presents the peak torque variations with respect to rotor field current. Thus, at full rotor field, saturation in the machine rotor (and to some extent the stator) negates the salient torque contribution.

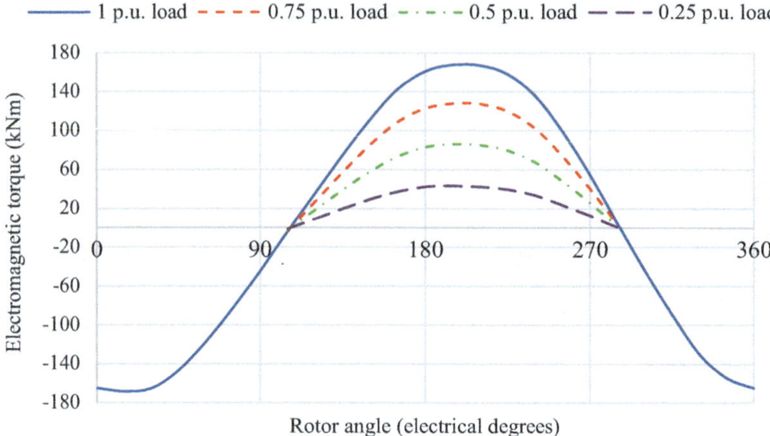

Fig. 4.25 Benchmark SG total torque at full-field and varying load

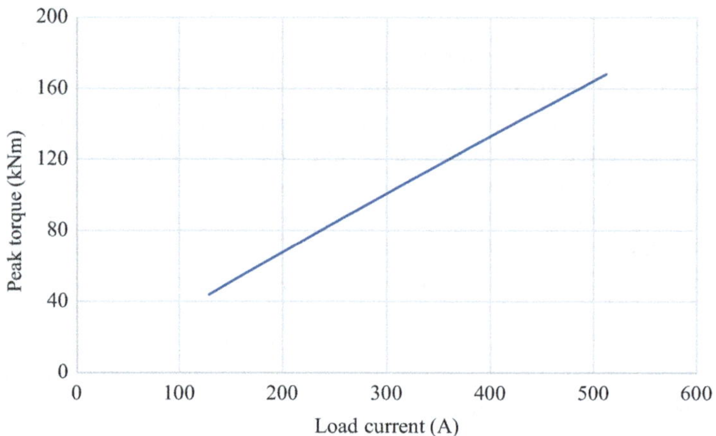

Fig. 4.26 Peak torque variation with respect to stator load current

4.3.6 Inductances

As discussed previously, the benchmark SG operates in saturation at rated full-field current and full-load stator current. One of the saturation effects is reduced machine inductance which contributes to the machine power capability. The SG synchronous reactance is a function of the rotor field current once the machine is outside the linear (magnetic) operation region, as shown in Fig. 4.29. Machine output power has a

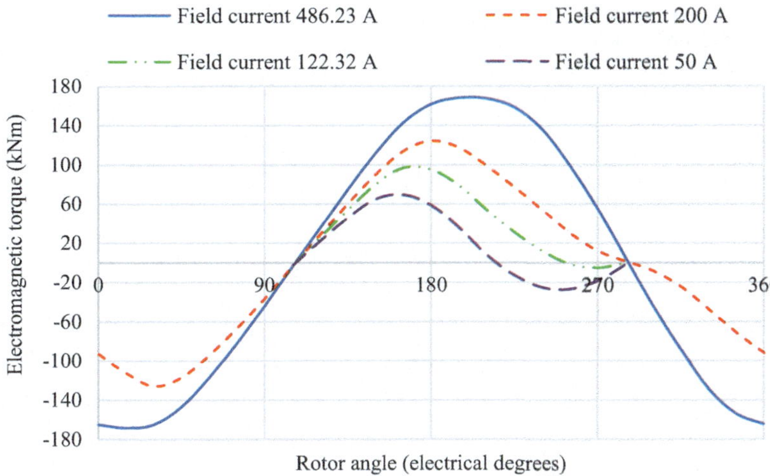

Fig. 4.27 Benchmark SG total torque at full-load stator current and varying rotor field current

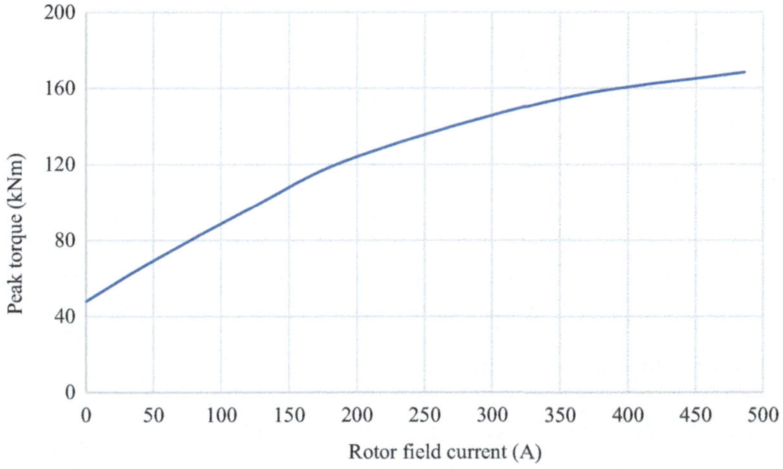

Fig. 4.28 Benchmark SG total torque at full-load stator current and varying rotor field current

reciprocal relationship with the machine reactance; hence, the machine output power increases as the machine goes into saturation as result of reduced inductance. Figure 4.30 presents the per-phase unsaturated self-inductance for the benchmark SG. The inductance is a function of rotor angle due to the rotor saliency, as discussed.

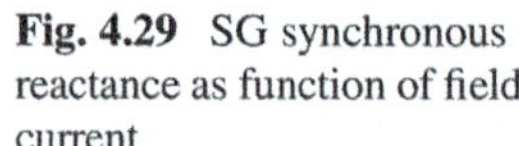

Fig. 4.29 SG synchronous reactance as function of field current

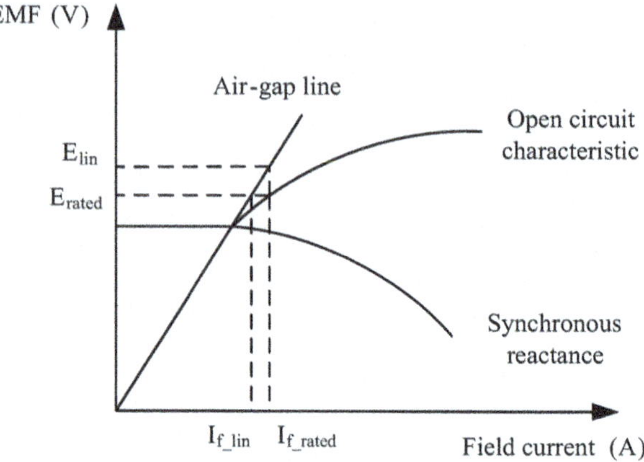

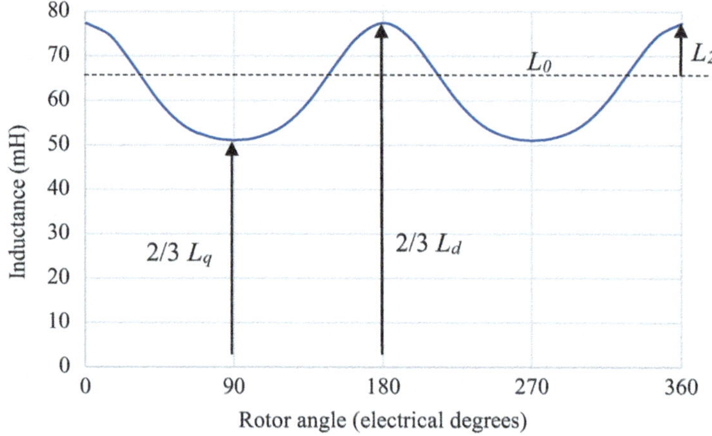

Fig. 4.30 Unsaturated phase self-inductance for the benchmark SG

The machine phase inductances are expressed as follows:

$$L_a = L_0 + L_2 \, \cos\left(2\theta_e\right)$$
$$L_b = L_0 + L_2 \, \cos\left(2\theta_e + \frac{2\pi}{3}\right)$$
$$L_c = L_0 + L_2 \, \cos\left(2\theta_e - \frac{2\pi}{3}\right)$$

$$(4.7)$$

and the mutual coupling between phases are:

$$M_{ab} = -\frac{1}{2} L_0 + L_2 \, \cos\left(2\theta_e - \frac{2\pi}{3}\right)$$

$$M_{ac} = -\frac{1}{2} L_0 + L_2 \, \cos\left(2\theta_e + \frac{2\pi}{3}\right) \tag{4.8}$$

$$M_{bc} = -\frac{1}{2} L_0 + L_2 \, \cos\left(2\theta_e\right)$$

where

L_a, L_b, L_c : Phase inductances for phase A, B, and C, respectively
M_{ab}, M_{ac}, M_{bc} : Mutual inductances between phases
L_0: Average phase self-inductance (including leakage)
L_2: Phase inductance amplitude variations above average

The d–q axis inductances shown in Fig. 4.30 are defined as follows:

$$L_d = \frac{3}{2} \left(L_0 + L_2\right)$$

$$L_q = \frac{3}{2} \left(L_0 - L_2\right) \tag{4.9}$$

The variation of inductances with rotor angle for a cylindrical rotor SG is generally neglected, and hence the term with angular variation, that is, L_2, is assumed zero. Hence, the machine synchronous inductance for a cylindrical SG is calculated as follows:

$$L_s = L_d = L_q = \frac{3}{2} L_0 \tag{4.10}$$

Note, when leakage inductance is neglected, the synchronous inductance for a cylindrical SG with an identical sinusoidally distributed winding can be calculated as follows:

$$L_s = \frac{m}{2} L_0 \tag{4.11}$$

where

m: Number of phases

To account for saturation in subsequent modeling, a saturation factor is calculated using the machine open-circuit characteristics. Referring to Fig. 4.29, the saturation factor is defined as:

Table 4.8 Benchmark SG inductances and resistance

Parameter	Value
L_0 (mH)	64.42
L_2 (mH)	13.22
L_d (mH)	116.45
L_q (mH)	76.8
L_{su} (mH)	96.63 (1 p.u.)
L_{ss} (mH)	42.76 (0.44 p.u.)
k_s	2.26
Phase resistance at 75 °C (Ω)	0.102

$$k_s = \frac{I_{f\_\mathrm{rated}}}{I_{f\_\mathrm{lin}}} = \frac{E_{\mathrm{lin}}}{E_{\mathrm{rated}}} \tag{4.12}$$

where

I_{f_rated}: Field current at rated stator load current (produces rated back-EMF)
I_{f_lin}: Field current that produces rated back-EMF on the linear air-gap characteristic
E_{lin}: Back-EMF produced by rated rotor field current on the air-gap linear characteristic
E_{rated}: Rated back-EMF

The saturated inductance is obtained by dividing the unsaturated inductance by a saturation factor:

$$L_{\mathrm{ss}} = \frac{L_{\mathrm{su}}}{k_s} \tag{4.13}$$

where

L_{ss}: Saturated inductance
L_{su}: Unsaturated inductance

Machine inductances for the benchmark SG are listed in Table 4.8 along with the machine per-phase resistance quoted at an operational steady-state winding temperature of 75 °C.

4.3.7 Modeling of Benchmark SG

The finite element parameters obtained for the benchmark SG are input to a MATLAB/Simulink simulation model of the machine. The benchmark SG is an industrial HV generator that is designed to connect to a three-phase, 50 Hz grid as shown in Fig. 4.31. Hence, the grid is modeled as an 11 kV line-to-line infinite bus-bar. The MATLAB model for the benchmark SG is shown in Fig. 4.32. The machine is analyzed with saturated and unsaturated inductance for result

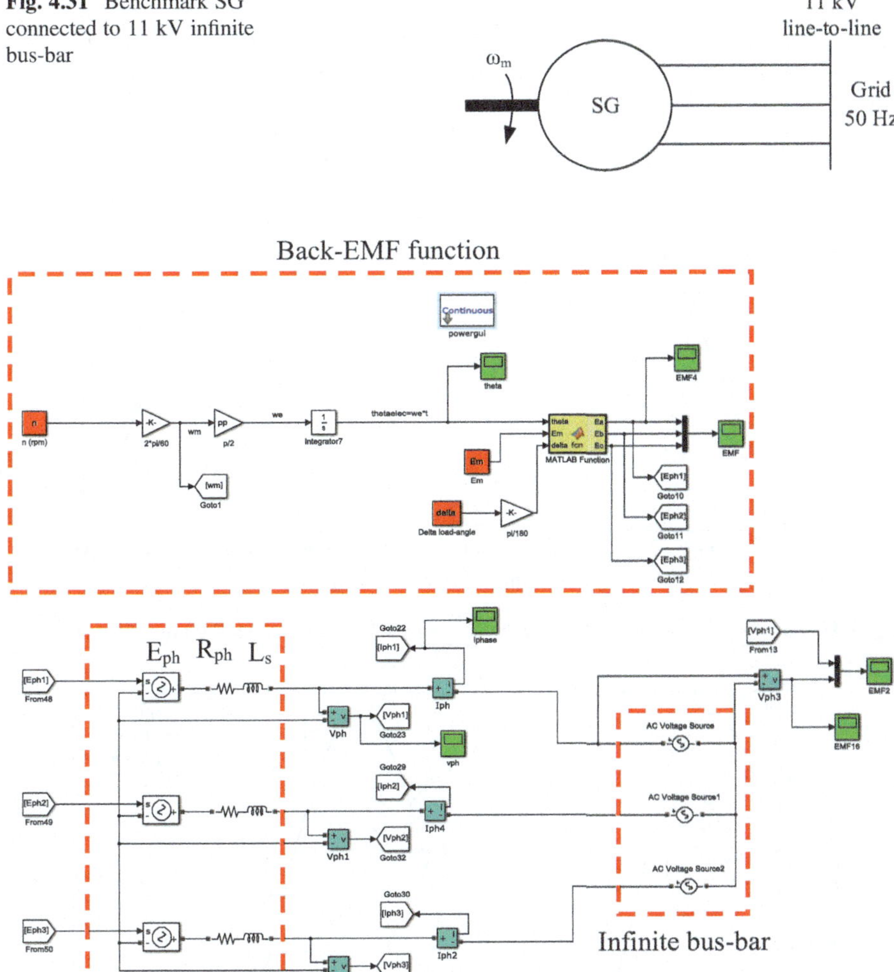

Fig. 4.31 Benchmark SG connected to 11 kV infinite bus-bar

Phase back-EMF in series
with machine impedance

Fig. 4.32 MATLAB model for benchmark SG connected to 11 kV infinite bus-bar

comparison. The unsaturated inductance is regarded here as 1 p.u.; hence, the saturated inductance is 0.44 p.u.

The SG output power is related to the sine of the load angle. Figure 4.33 shows power-angle characteristics for the benchmark SG. As seen, the power increases as the inductance reduces. Therefore, a machine operating in the saturation region has a better power capability due to reduced inductance. Machine phase back-EMF, phase voltage, and current for a load angle of 22.15 electrical degrees are plotted in Fig. 4.34. In generator mode of operation, the back-EMF leads the phase voltage

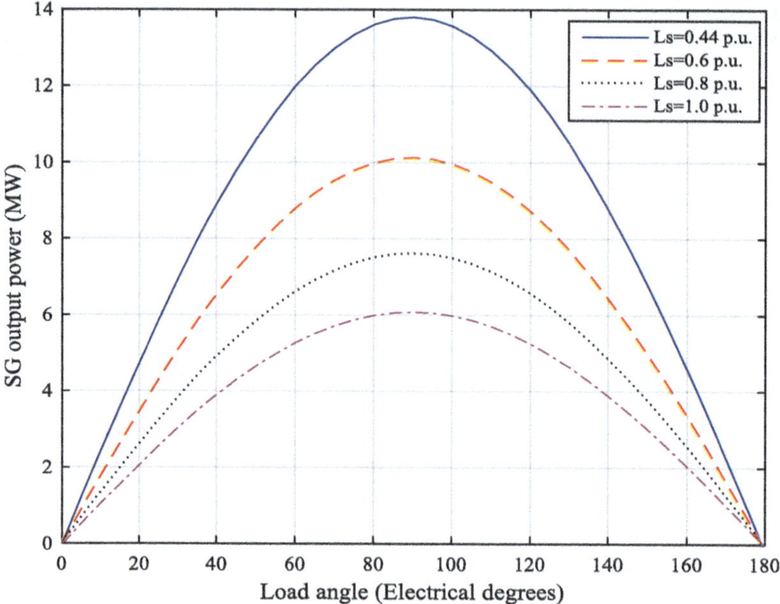

Fig. 4.33 Power-angle characteristics for the benchmark SG connected to an infinite bus-bar

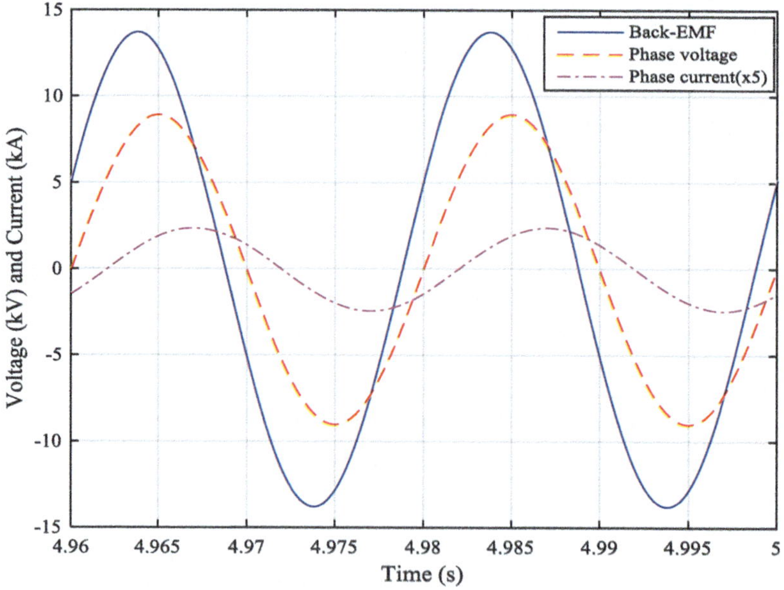

Fig. 4.34 Benchmark SG back-EMF, phase voltage, and current

Table 4.9 Benchmark SG simulated results

Parameter	Simulated
Rated output power (MW)	5.52
Full-field rotor current (at rated power) (A)	486.23
Open-circuit phase RMS back-EMF at full-field current (kV)	10.32
Grid line-to-line RMS voltage (kV)	11
Full-load stator RMS current (A)	357.2
Load angle (degrees)	22.15
Power factor at full-load stator current (p.u.)	0.81
Grid frequency (Hz)	50
Synchronous inductance (Ω)	13.43
Phase resistance (Ω) at 75 °C	0.102
Rotor winding resistance (Ω) at 75 °C	0.13

and power is injected to the bus-bar. The phase current lags the phase voltage by 35.57°, that is, a power factor of 0.81. Table 4.9 shows simulation results for the benchmark SG.

4.4   Summary

In this chapter, different machine topologies and power conversion schemes for wind generation systems are reviewed. The hybrid generator concept is discussed, and a three-phase 11 kV, 5.52 MW synchronous generator is chosen as benchmark machine based on which the high-voltage hybrid generator is designed in Chap. 5. Benchmark SG air-gap flux-density, flux-linkage, back-EMF, torque, and inductance are evaluated using finite element analysis. It is shown that machine operates in saturation region at full-load stator current and full-field rotor current. This results in a reduced inductance and improved power capability.

Chapter 5
Multiphase Machine Design

5.1 Introduction

As discussed in preceding chapters, the HG has two rotor parts, a WF and a PM. Both rotors are mounted on the same shaft; hence, they rotate with the same speed and share one stator. The HG WF and HG PM rotors can be designed as separate machines that are then combined with an appropriate split ratio to form an HG. The HG stator is the same as the benchmark SG, that, the same number of slots and geometry. However, two options will be considered for the stator winding:

- A 3-phase winding, as per the benchmark SG
- A 3-phase winding within the same stator slot and geometry

The 3- and 9-phase designs will be compared to illustrate the benefits of the multiphase solutions. The rotor geometry for the WF is the same as the benchmark SG rotor with 10 salient poles. The rotor geometry for the PM section is also 10 poles with suitable magnet dimensions. The design will consider a full WF and a full PM solutions and an HG having a WF:PM split ratio.

5.2 HG Design Philosophies

In Chap. 2, the wind turbine operation was divided into two regions:

 (i) Limited speed region of operation
(ii) High speed region of operation

For the limited speed region, the HG speed is rated at 1 p.u., Fig. 2.15, while HG speed in the high speed region could be as high as 1.67–2.66 p.u. ($\approx$3 p.u.), Fig. 2.17, depending on the pitch angle. Therefore, design of the HG varies depending on the choice of wind turbine operational region. In this section, different

© Springer Nature Switzerland AG 2020
O. Beik, A. S. Al-Adsani, *DC Wind Generation Systems*,
https://doi.org/10.1007/978-3-030-39346-5_5

design philosophies for the HG in both limited and high speed regions are investigated to address the wind turbine and system requirements. However, as it was discussed in Chap. 2, the high speed region of operation is not an industry practice for the wind turbine generator. Hence, the remaining section of this chapter only considers HG design for limited speed operation of the wind turbine.

5.2.1 HG Design Philosophy for Limited Speed Region

In a wind farm, the wind turbines experience varying wind velocity with time. This is due to uneven distribution of the wind mass, uncertainty of the wind direction, and also the location of the wind turbines. Therefore, in a large farm such as Walney, the output voltage and power of the wind turbine generators will vary slightly from one another [14, 60–65]. As discussed previously, for systems typical of the Walney wind farm, the two back-to-back VSCs decouple the generator from the grid and control the voltage and frequency to match that of grid. However, in the HG system, there are no VSCs; instead, the HG is connected to a passive rectifier that has no inherent control functionality. Therefore, the WF of the HG must control the output voltage. Since the HG rectifiers for all the turbines in the farm are electrically connected in parallel, any voltage disparity between generators will cause the rectification stage potentially to turn off. Therefore, the WF must be designed for the largest wind velocity variations between the wind turbines in the wind farm. A maximum of less than 20% wind velocity variation between wind turbines throughout a wind farm is reported in the literature and measured in the industrial settings. Here, a maximum wind velocity variation of 25% is considered for the HG design to assure some margin in design.

The HG WF design is dependent on voltage and hence speed variations. Therefore, the split ratio between PM and WF rotor sections of the HG can be considered as varying from a fully WF rotor or traditional SG to some ratio of PM to WF excitation. To satisfy the 25% wind velocity variation, an HG with a PM:WF ratio of 0.75:0.25 is considered and compared with a (100%) WF (or SG). Here, the 100% WF (or SG) machine is considered to be a reference 1 p.u. machine.

First consider a fully WF (or SG) machine. Figure 5.1 presents total terminal voltage of a SG against rotor speed, while the wind turbine is operating within the limited speed operation region. The output voltage of the SG is considered to be proportional to the rotor speed since the rotor WF current is fixed. As discussed in Chap. 2, Fig. 2.15, there are some speed and power perturbations around the rotor base speed which explains the voltage and current perturbations around base speed of Fig. 5.1.

Now consider that the SG is scaled down to act as the WF rotor of the HG. Therefore, the WF rotor provides 25% of the maximum HG terminal voltage and the PM rotor induces the remaining 75%. The HG total terminal voltage versus rotor speed variation is depicted in Fig. 5.2 when the HG is operating in the limited speed range. This WF boost voltage ensures that the HG is in a safe operating

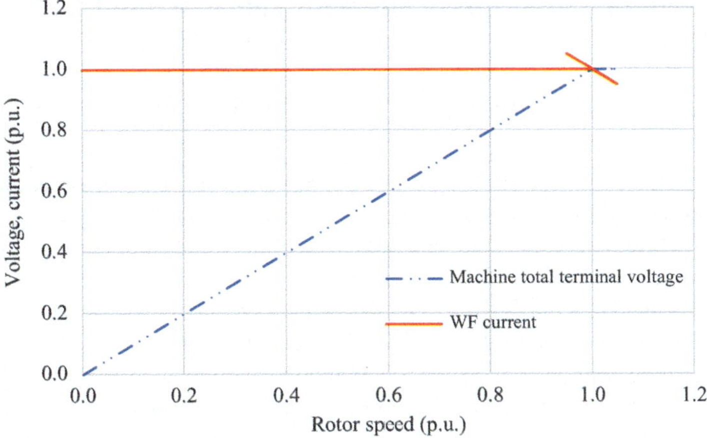

Fig. 5.1 Voltage variation of a fully WF (SG) for a turbine operating in the limited speed region

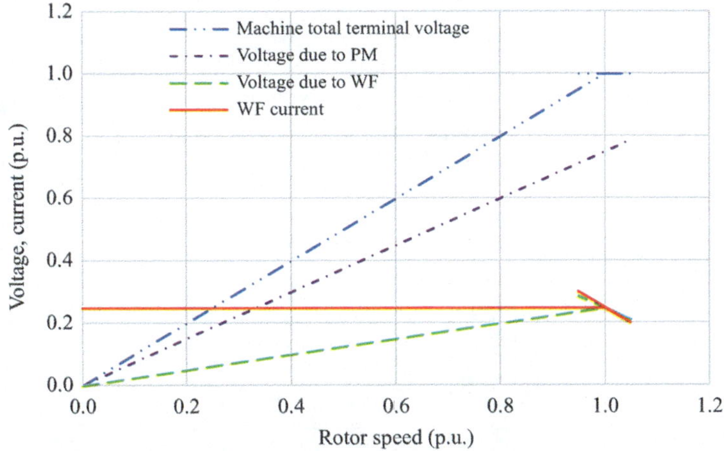

Fig. 5.2 Voltage variations for an HG with 75% induced voltage due to PM rotor and 25% induced voltage due to WF rotor for a turbine operating in the limited speed region

condition in the event of loss of rotor field, as opposed to a design where the WF voltage "bucked" (reduced) that of the PM field. Loss of rotor field also acts as a protection feature in the event of generator-faulted operation.

5.3   Design of Three-Phase HG with 100% Wound Field (WF)

5.3.1   Generator Connected to a Passive Rectifier

The three-phase HG with 100% WF is essentially the same machine as the benchmark SG presented in Chap. 4. However, the benchmark SG is connected to a three-phase grid while here connected to a passive rectifier. The SG output power varies with the rotor angle when the machine is connected to an infinite bus-bar grid. However, for the HG connected to a rectifier, the rotor angle variation is constrained by the action of the diode bridge rectifier; hence, some performance assessment has to be performed. Figure 5.3 shows a schematic diagram of the 3-phase HG with 100% WF, referred to as 3-phase HG WF, connected to a passive rectifier.

A MATLAB model is set up for the 3-phase HG WF to evaluate the machine performance, as shown in Fig. 5.4. The DC link is connected to a DC power supply (battery) acting as the load. The model is set up to simulate an HG WF connected to rectifier and also a voltage source converter (VSC) so that the machine performance can be evaluated when connected to an uncontrolled passive rectifier and to a fully

Fig. 5.3 A 3-phase HG WF connected to a passive rectifier

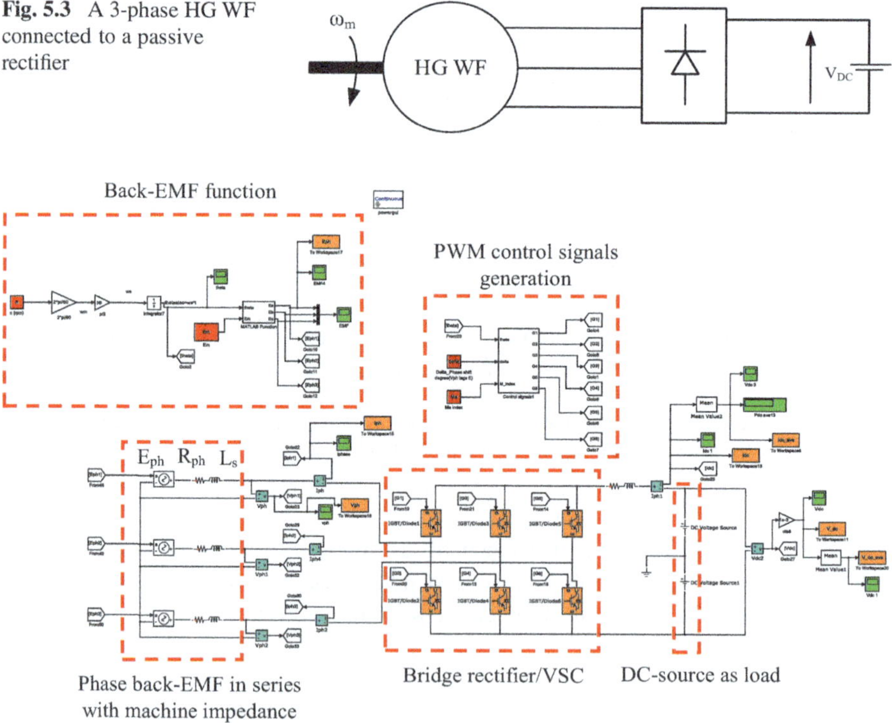

Fig. 5.4 MATLAB model for the 3-phase HG WF connected to a rectifier or VSC

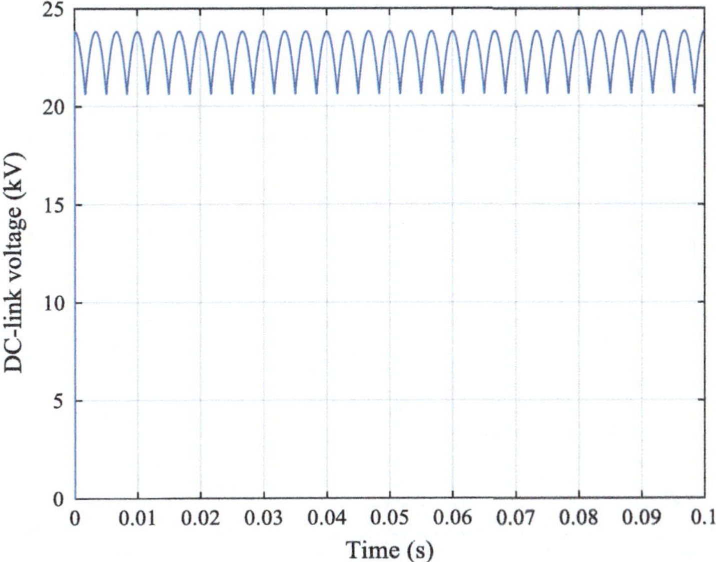

Fig. 5.5 No-load DC-link voltage at the rectifier output

Table 5.1 DC-link and machine voltages for the 3-phase HG WF connected to a passive rectifier

Parameter	Value (kV)
Phase peak voltage	13.77
Line-to-line RMS voltage	16.86
Average no-load DC-link voltage	22.77
Peak no-load DC-link voltage	23.85
Minimum no-load DC-link voltage	20.65
No-load DC-link voltage variations	3.2

controlled VSC and the two schemes compared. For passive rectifier operation, the gates of the IGBTs in the model are disabled. Therefore, the converter acts as a rectifier using the parallel diodes of the IGBTSs. The average of the rectified DC-link voltage can be analytically calculated as follows:

$$V_{dc}^{ave} = \frac{3}{\pi} \sqrt{2}\, V_{LL}^{rms} \qquad (5.1)$$

where

V_{LL}^{rms}: RMS line-to-line voltage

The no-load DC-link voltage for the rectifier output is plotted in Fig. 5.5. The average of the simulated no-load DC link is 22.77 kV, which is consistent with the analytical value calculated from Eq. (5.1). Table 5.1 lists the DC-link and HG WF main voltages.

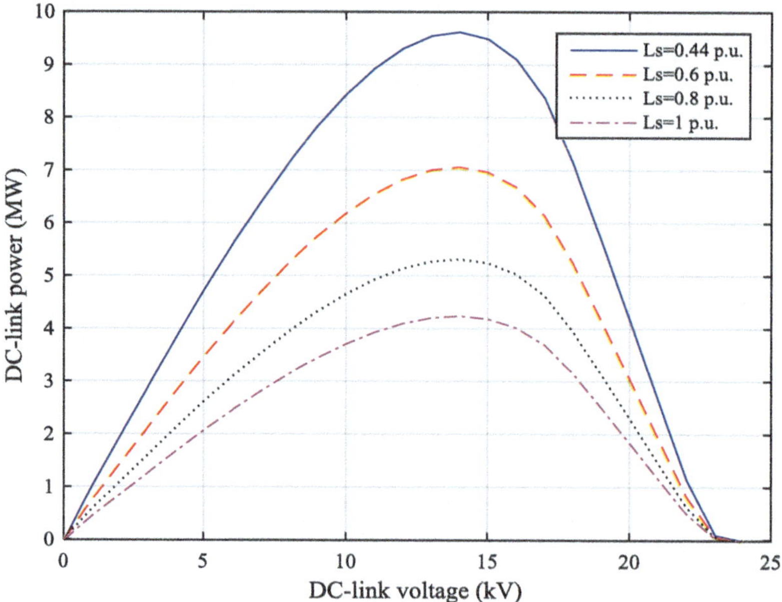

Fig. 5.6 DC-link power versus DC-link voltage variations at the rectifier output

Figure 5.6 shows the DC-link power with respect to DC-link voltage variations
for different inductances when the 3-phase HG WF is connected to passive rectifier.
The phase back-EMF, phase voltage and current for a DC-link voltage of 14 kV, and
machine inductance of 0.44 p.u. are plotted in Fig. 5.7. At a DC-link voltage of
14 kV, the fundamental component of the phase current lags the fundamental of the
phase voltage by 5.12°, that is, a power factor of 0.996 for the fundamental
components. The phase current is defined by the load, that is, the battery voltage.
Therefore, the power factor is defined by the load and cannot be controlled via
passive diode rectifier. The phase back-EMF is a sine wave generated by the HG
WF, but the phase voltage is a 6-step quasi-square waveform with $\pm 2/3\ V_{DC}$ and $\pm 1/3\ V_{DC}$ voltage levels (Fig. 5.7). The instantaneous DC-link power and current are
plotted in Fig. 5.8. The DC-link power perturbations are due to the ripples in the
DC-link current.

5.3.2 Generator Connected to a VSC

In the previous section, performance of the 3-phase HG WF connected to a rectifier
was discussed. In commercial wind turbines, the Walney wind turbines, for instance,
the generator is sometimes connected to an active power electronic converter, that is,
VSC. The VSC allows vector control of the HG WF, hence improving the output
power capability of the machine. In this section, the performance of the HG WF

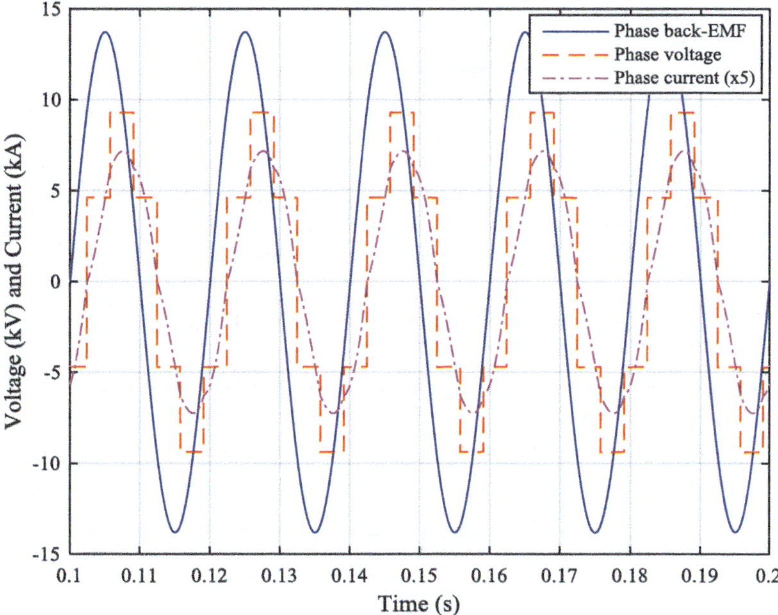

Fig. 5.7 Phase back-EMF, phase voltage, and current for a DC-link voltage of 14 kV and 0.44 p.u. inductance for a 3-phase HG WF connected to a passive rectifier

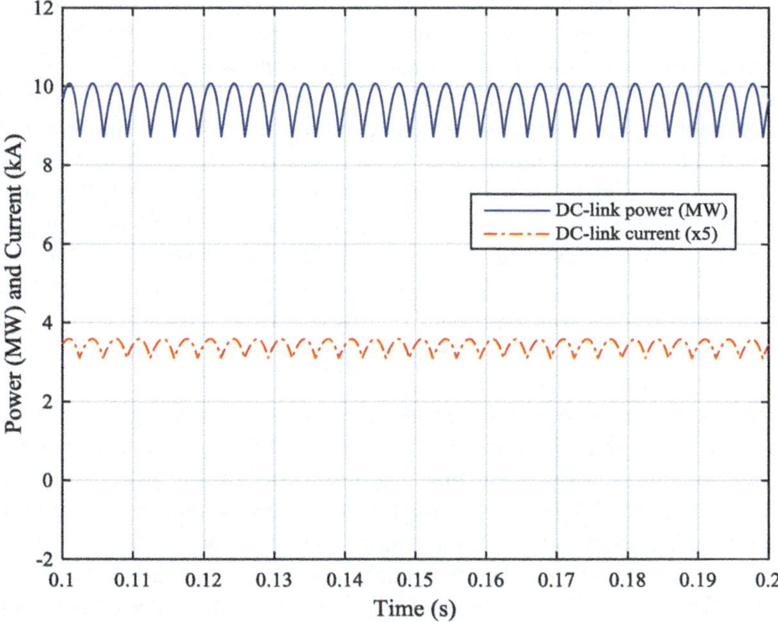

Fig. 5.8 DC-link power and current for a DC-link voltage of 14 kV and 0.44 p.u. inductance for a 3-phase HG WF connected to a passive rectifier

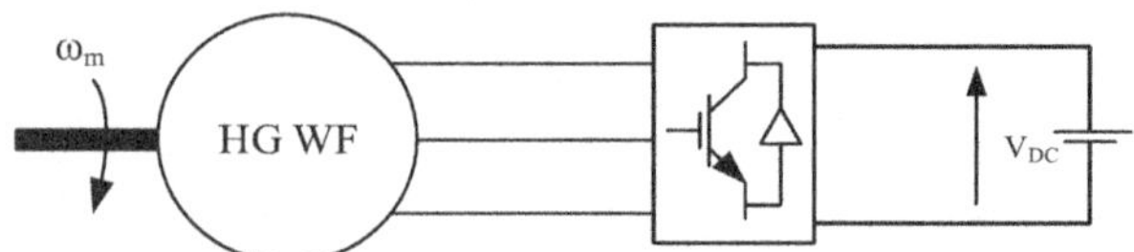

Fig. 5.9 Three-phase stator HG WF connected to a VSC

connected to a three-phase VSC with two control schemes, PWM control and overmodulated, is evaluated and results are compared with that of rectifier-connected HG WF. Figure 5.9 shows the three-phase HG WF connected to a VSC and a battery as load.

The IGBT's gate signals shown in the MATLAB model in Fig. 5.4 are generated via comparing a 3 kHz sawtooth carrier signal with a sine wave control (modulating) signal. The phase fundamental component is synthesized based on the switching sequences and hence will be proportional to the control signal, that is, a sine waveform. An amplitude modulation index is defined as follows:

$$m_a = \frac{\widehat{V}_{ctrl}}{\widehat{V}_{saw\text{-}tooth}} \tag{5.2}$$

where

$\widehat{V}_{ctrl}$: Control signal amplitude

$\widehat{V}_{saw\text{-}tooth}$: Carrier signal amplitude

The amplitude of the phase voltage fundamental component is controlled via the modulation index. Moreover, the phase angle and frequency of the phase voltage fundamental component are proportional to the control signal phase angle and frequency. Therefore, amplitude, angle, and frequency of the phase voltage fundamental component are controlled by the control signal. For a sinusoidal PWM control scheme, the modulation index varies between 0 and 1, and the output phase voltage envelope is a 6-step quasi-square wave with $\pm 2/3\ V_{DC}$ and $\pm 1/3\ V_{DC}$ voltage levels. Figure 5.10 shows phase back-EMF, phase voltage, and phase current of the 3-phase HG WF when connected to a VSC with PWM control, modulation index of 0.95, DC-link voltage of 14 kV, and for an angle of 60° between the fundamental voltage component and the back-EMF, i.e., the load angle. For this load angle, the phase current lags the fundamental component of the phase voltage by 0.11°, that is, a power factor of ≈1.0 p.u.

DC-link power and current for the same conditions stated in Fig. 5.10 are plotted in Fig. 5.11. The average power and current in the DC link are constant, while the instantaneous values vary with time as result of PWM switching. For the overmodulated control scheme, the modulation index is greater than 1.0. For a modulation index from 0 to 1.0, the phase voltage increases linearly. Above a modulation index of 1.0, the phase voltage increases nonlinearly to a maximum value of 1.273 p.u. as shown in Fig. 5.12. The machine phase back-EMF, voltage, and current for overmodulated control scheme for the same conditions as PWM

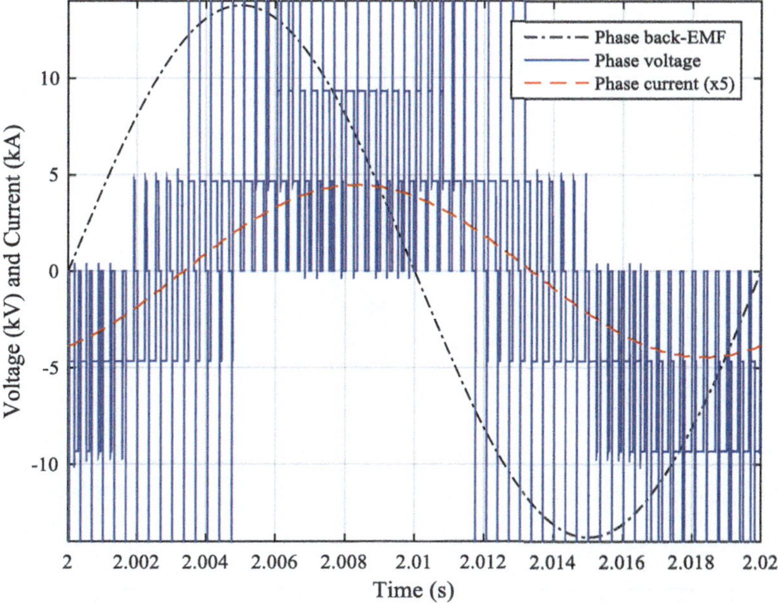

Fig. 5.10 Phase back-EMF, voltage, and current for a 3-phase HG WF connected to a VSC with PWM control, modulation index 0.95, DC-link voltage 14 kV, inductance 0.44 p.u., and load angle 60°

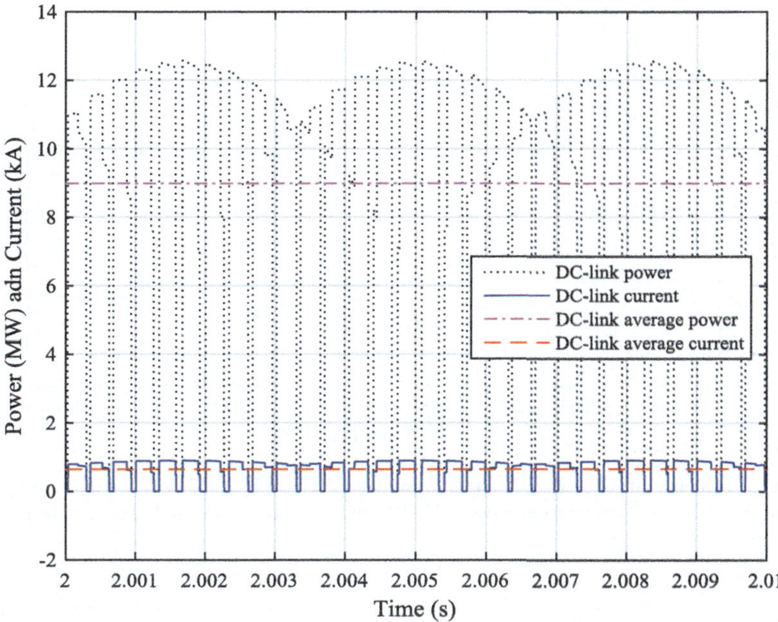

Fig. 5.11 DC-link power and current for the same conditions as in Fig. 5.10

Fig. 5.12 Amplitude of the phase voltage fundamental component variations with respect to modulation index

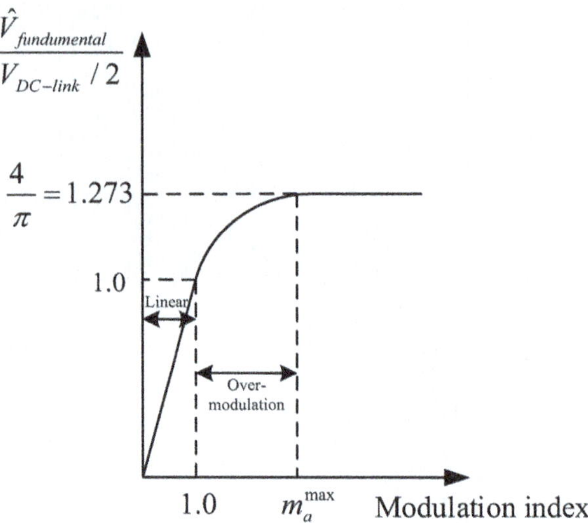

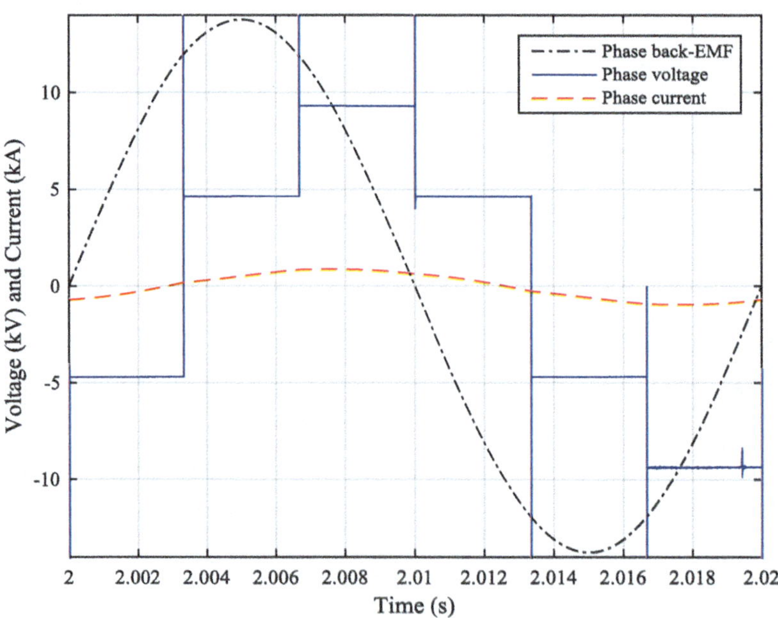

Fig. 5.13 Phase back-EMF, voltage, and current for a 3-phase HG WF connected to a VSC with overmodulated control, modulation index 100, DC-link voltage 14 kV, inductance 0.44 p.u., and load angle 60°

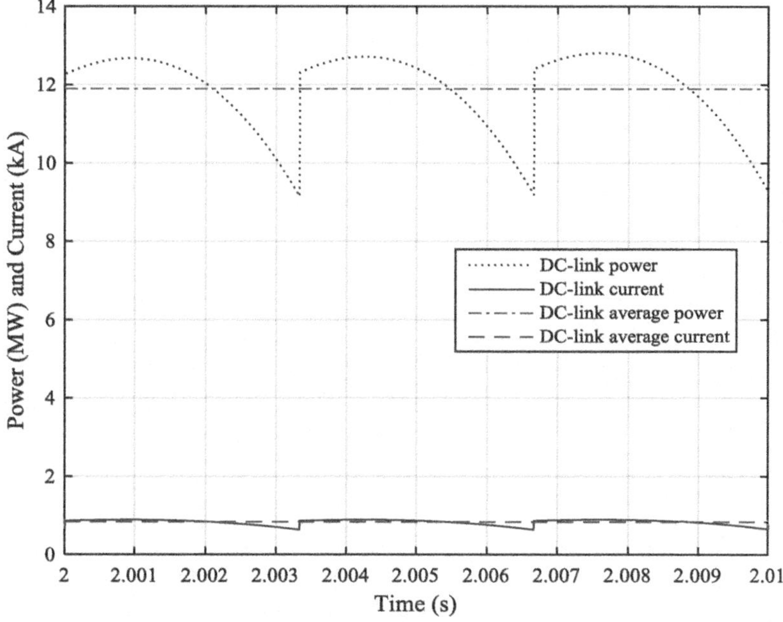

Fig. 5.14 DC-link power and current for the same conditions as in Fig. 5.13

control scheme but with a modulation index of 100 are plotted in Fig. 5.13. The phase voltage is a 6-step square wave with $\pm 2/3\ V_{DC}$ and $\pm 1/3\ V_{DC}$ voltage levels.

In the overmodulated scheme, the phase current lags $10.06°$ behind the phase voltage fundamental component, that is, a 0.98 power factor. Note that the load angle in both PWM and overmodulated is kept at $60°$. The DC-link power and current for the overmodulated control scheme are shown in Fig. 5.14. The DC-link average power for the overmodulated VSC control is 11.91 MW while for the PWM control scheme and the same load angle is 8.98 MW. This is due to a maximum 0.95 modulation index for the PWM control scheme (to maintain sinusoidal phase current control) and also the 1.273 factor increase in phase voltage for the overmodulated case. Figure 5.15 illustrates the DC-link power with respect to load angle for PWM and overmodulated control schemes. It is seen that the power gain for the overmodulated control scheme occurs at all load angles. However, in the overmodulated scheme, the amplitude of the phase voltage fundamental component cannot be controlled. The switching frequency for the overmodulated control is low which makes it desirable for high-power applications.

Table 5.2 shows the maximum DC-link power for a 3-phase HG WF connected to a passive rectifier and VSC. The DC-link power in the rectifier case is lower due to passive rectifier inability to control load angle, whereas with a VSC, the load angle can be controlled to $90°$. Therefore, when a passive rectifier replaces a VSC (or active rectifier), then in order to get the same DC-link power, the 3-phase HG

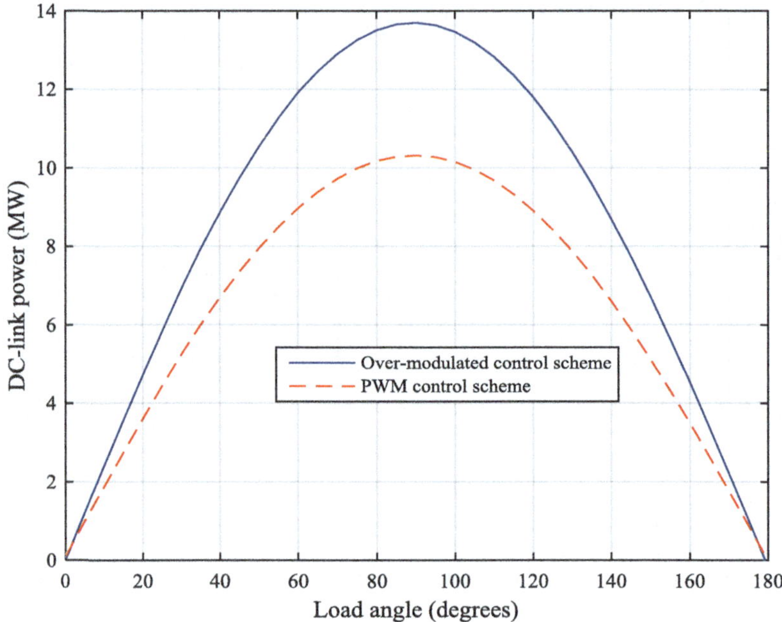

Fig. 5.15 DC-link power with respect to load angle for PWM and overmodulated control for a three-phase HG WF connected to a VSC, DC-link voltage 14 kV, and inductance 0.44 p.u

Table 5.2 Maximum DC-link power for the 3-phase HG WF connected to a passive rectifier and VSC

Three-phase HG WF connected to	Maximum DC-link power (MW)	DC-link power ratio (VSC/rectifier)	
		Overmodulated	PWM
Passive rectifier	9.63	1.42	1.1
VSC with overmodulated control scheme	13.70		
VSC with PWM control scheme	10.31		

WF needs to be redesigned for a factor of 1.42 and 1.1 higher power for overmodulated and PWM schemes, respectively.

Hence, for the same application and DC-link power, it can be concluded that the machine rating must be larger if interfaced by a passive rectifier than when interfacing by a VSC.

5.4 Design of 9-Phase HG with 100% Wound Field (WF)

In this section, the 3-phase HG with 100% WF, that is, 3-phase HG WF, is rewound as a 9-phase stator winding, referred to as 9-phase HG WF. The machine rotor and stator geometries, number of poles (10), number of stator slots (135), and physical specifications are maintained the same for the purpose of comparison. For the 9-phase winding, the phase band from Eq. (4.1) is 1.5. The winding is short-pitched and also fractional-slot. Therefore, in order to rewind the stator to include 9 phases, a pattern of 2 and 1 coils is required. Figure 5.16 illustrates a pattern for the 9-phase stator winding for the HG WF. The two-set and one-set coils are connected in series to form a pattern. The pattern is repeated five times; therefore, there are 15 coils per phase for the 9-phase winding. The coil span is 12 slots and the winding is double layer. For the 3-phase winding, each coil had 4 turns (conductors), but for the 9-phase winding, each coil has 12 turns. Therefore, having the same coil span, the number of turns per phase for the 3- and 9-phase windings is 180: for 3-phase, 45 coils and 4 turns per coil, while for the 9-phase, 15 coils and 12 turns per coil. The turn cross-sectional area in the 9-phase winding is one-third that of the turn cross-sectional area in the 3-phase winding. Therefore, the per-phase resistance of the 9-phase winding is three times that of the 3-phase winding. Based on the value listed in Table 4.10 for the per-phase resistance of the 3-phase winding, the per-phase

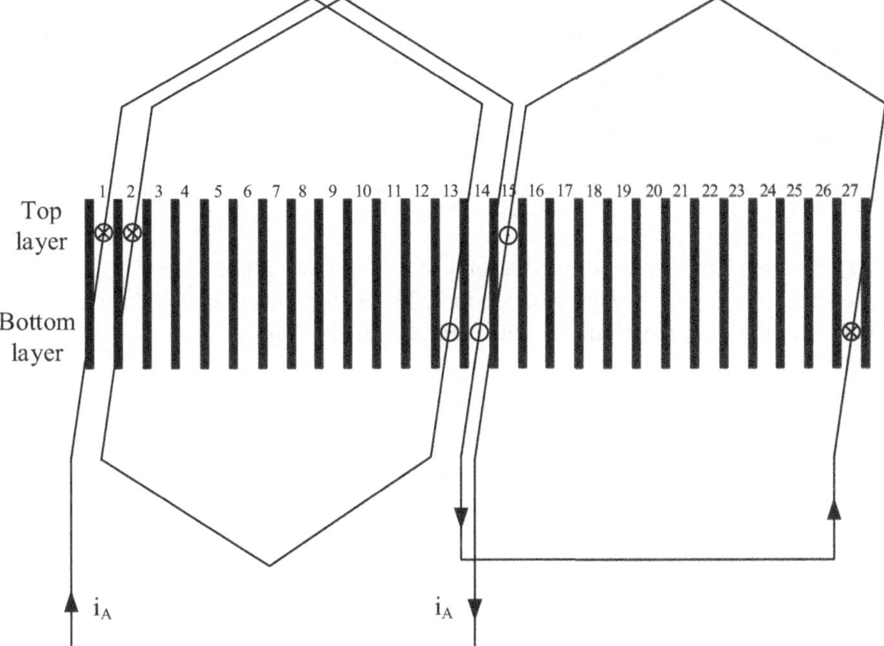

Fig. 5.16 Fractional-slot, short-pitched stator winding for the 9-phase HG WF

resistance for the 9-phase winding at the operational temperature of 75 °C is 0.305 Ω. Note that although the phase resistance is three times higher for the 9-phase design, the copper loss for the 3- and 9-phase winding schemes is equitable since the 9-phase currents are reduced by a factor of 3.

For the purpose of verification, the per-phase resistance of the 9-phase winding is also calculated from the geometry of the machine as follows. The turn resistance is:

$$R_{\text{turn}}^{9\phi} = \frac{\rho_{\text{cu}}\, l_{\text{turn}}}{A_{\text{turn}}} \tag{5.3}$$

and

$$l_{\text{turn}} = 2\, l_a + 4\, \sqrt{2}\, \frac{(\text{coil pitch})}{2} \tag{5.4}$$

where

ρ_{cu}: Copper electrical resistivity (Ω.m)
l_{turn}: Length of a turn in a coil (mm)
A_{turn}: Cross section of a turn in a coil (mm^2)
l_a: Axial length (mm)

Each turn in a coil has two sides plus end winding. The two sides have the same length as the machine axial length. According to Fig. 5.16, the end winding has a diamond (triangular) shape whose base length is the same as the coil pitch and height is half of the coil pitch, where the coil pitch is 12 slots. There are two such triangles, one at each end of the machine.

Therefore, for an N-turn coil:

$$R_{\text{coil}}^{9\phi} = N_{\text{stator}}\, R_{\text{turn}}^{9\phi} \tag{5.5}$$

There are 12 turns per coil and a total of 15 coils per phase. These 15 coils are connected in series using 9 interconnections. The interconnection also has a diamond shape with the base length of 13 slots and a height 13/2 slots:

$$l_{\text{interconnection}} = 2\, \sqrt{2}\, \frac{(13\ \text{slots})}{2} \tag{5.6}$$

$$R_{\text{interconnection}}^{9\phi} = \frac{\rho_{\text{cu}}\, l_{\text{interconnection}}}{A_{\text{turn}}} \tag{5.7}$$

where

$l_{\text{interconnection}}$: Interconnection length (mm)
$R_{\text{interconnection}}^{9\phi}$: Interconnection resistance (Ω)

Therefore, the phase resistance is:

Table 5.3 Nine-phase winding specifications for the HG WF

Item	Value
Copper electrical resistivity (Ω.m)	1.68×10^{-8}
Turn length (mm)	2498.1
Turn cross-section area (mm^2)^a	31.72
Axial length (mm)	400
Coil pitch (12 slots) (mm)	600.36
Turn resistance (Ω)	0.00137
Coil resistance (Ω)	0.0165
Interconnection length (mm)	919.79
Interconnection resistance (Ω)	0.0005
Phase resistance (Ω) at 20 °C	0.252
Temperature coefficient of resistance (°C^{-1})	0.003862
Phase resistance (Ω) at 75 °C	0.306

^{a}Turn cross section for 3-phase winding (with 6 strands) = 95.16 mm^2 ($= 6.1 \times 2.6 \times 6$)

$$R_{\text{ph}}^{9\phi} = 15\, R_{\text{coil}}^{9\phi} + 9\, R_{\text{interconnection}}^{9\phi} \tag{5.8}$$

Table 5.3 lists the winding specifications. Note that the calculated resistances are at 20 °C. The resistance at the operational temperature of the machine, 75 °C, is given by:

$$R_{\text{ph}(T_2)}^{9\phi} = R_{\text{ph}(T_1)}^{9\phi} \left[1 + \alpha_T \left(T_2 - T_1\right)\right] \tag{5.9}$$

where

T_2: Machine operational temperature 75 °C
T_1: Room temperature 20 °C
α_T: Temperature coefficient of resistance °C^{-1}

The phase resistance calculated for the 9-phase winding from the machine geometry is the same as that estimated using the 3-phase resistance confirming the accuracy of the method.

The rotor WF resistance is calculated from the rotor and coil geometry. Rotor winding has 10 coils, each wound around the 10 salient rotor poles and connected in series. Each rotor coil has 51 turns. Each turn has two sides, each the same length as the machine active axial length (400 mm) plus end windings that have the same length as the width of the rotor poles (350 mm). Also, the length of the interconnection between the series coils is the same as the pole pitch. The turn resistance and length of a rotor coil turn is calculated via Eqs. (5.10) and (5.11):

Table 5.4 Rotor field winding specifications for the HG WF

Item	Value
Length per turn (mm)	1500
Turn cross-sectional area (mm^2)	112
Turn resistance (mΩ)	0.225
Number of turns per coil	51
Coil resistance (mΩ)	11
Winding interconnection length (mm)	675.44
Winding interconnection resistance (mΩ)	0.1
Resistance (Ω) at 20 °C	0.11
Resistance (Ω) at 75 °C	0.13

$$R_{\text{turn}}^{\text{rotor}} = \frac{\rho_{\text{cu}}\, l_{\text{turn}}^{\text{rotor}}}{A_{\text{turn}}^{\text{rotor}}} \tag{5.10}$$

$$l_{\text{turn}}^{\text{rotor}} = 2\, l_a + 2\, l_w^{\text{poles}} \tag{5.11}$$

Hence:

$$R_{\text{coil}}^{\text{rotor}} = N_{\text{rotor}}\, R_{\text{turn}}^{\text{rotor}} \tag{5.12}$$

$$R_{\text{interconnection}}^{\text{rotor}} = \frac{\rho_{\text{cu}}\, l_{\text{interconnection}}^{\text{rotor}}}{A_{\text{turn}}^{\text{rotor}}} \tag{5.13}$$

Therefore, the rotor resistance is:

$$R_{\text{rotor}} = 10\, R_{\text{coil}}^{\text{rotor}} + 9\, R_{\text{interconnection}}^{\text{rotor}} \tag{5.14}$$

where

$R_{\text{turn}}^{\text{rotor}}$: Turn resistance for the rotor coil (Ω)

$l_{\text{turn}}^{\text{rotor}}$: Length of a turn in a rotor coil (mm)

$A_{\text{turn}}^{\text{rotor}}$: Cross section of a turn in rotor coil (mm^2)

l_w^{poles}: Width of rotor pole (mm)

$R_{\text{coil}}^{\text{rotor}}$: Rotor coil resistance (Ω)

N_{rotor}: Number of turns in a rotor coil

$l_{\text{interconnection}}^{\text{rotor}}$: Length of rotor winding interconnection (mm)

$R_{\text{interconnection}}^{\text{rotor}}$: Rotor interconnection resistance (Ω)

Rotor winding specifications are listed in Table 5.4.

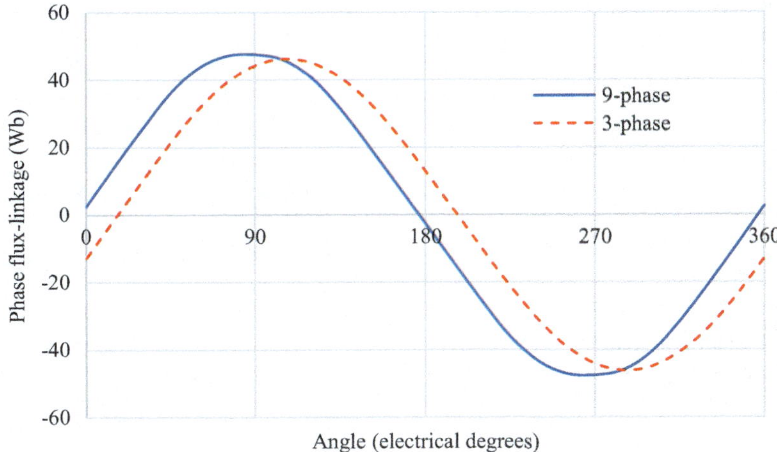

Fig. 5.17 Phase flux-linkage for the HG WF at a full-field current 486.23 A

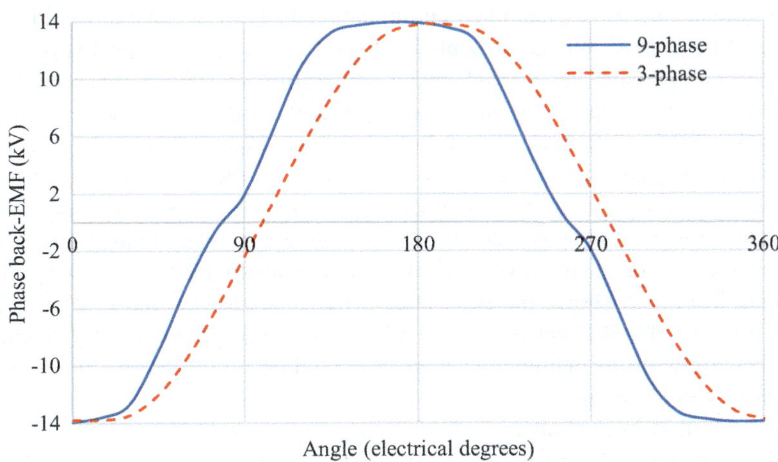

Fig. 5.18 Phase back-EMF for the HG WF at a full-field current 486.23 A

Table 5.5 RMS and peak voltages for 3- and 9-phase HG WF at full field of 486.23 A and 600 RPM

HG WF with	3-phase stator winding	9-phase stator winding	Voltage ratio (9-phase)/(3-phase)
Peak phase voltage (kV)	13.77	13.94	1.0123
RMS phase voltage (kV)	10.37	10.80	1.0415

Fig. 5.19 Phasor diagram
for 3-phase winding, 5-set
and 4-set pattern

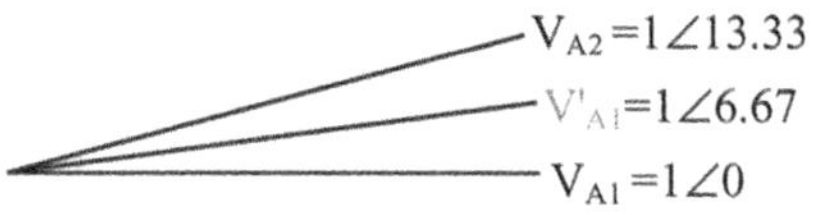

Fig. 5.20 Phasor diagram
for 9-phase winding, 2-set
and 1-set patterns

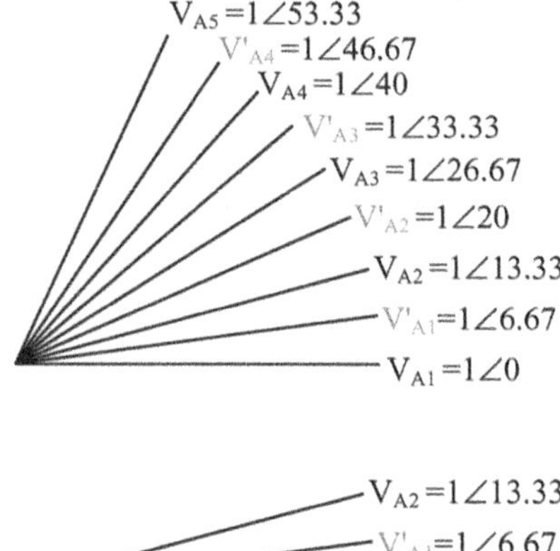

5.4.1   Comparison of 3-Phase and 9-Phase Machines

The flux-linkage per-phase and open-circuit induced back-EMF at a full-field current
of 486.23A for the HG WF are presented in Figs. 5.17 and 5.18, respectively. For the
purpose of comparison, the 3- and 9-phase waveforms are shown. It is seen that the
back-EMF for the 9-phase winding is less sinusoidal compared to that of the 3-phase
back-EMF. This is due to a lower number of coils per phase in the 9-phase winding.
Note that the angular difference (phase shift) between the two windings is simply
due to a shift in the coil center reference. Since the HG is connected to a passive
rectifier in the wind generation scheme, it is no longer necessary to have a sinusoidal
output voltage; indeed, a move toward a more trapezoidal waveform would be
beneficial. Table 5.5 shows the peak and RMS values for the induced back-EMF
for the 3-phase HG WF and 9-phase HG WF. It is seen from Table 5.5 that the
9-phase winding results in higher peak and RMS values for the phase voltage. This is
explained using the phasor diagram.

Assuming that individual coils have sinusoidal back-EMFs (which is not the
case), the winding coil distribution factor can be estimated from the phasor diagram
for a pattern of 5-set and 4-set coils for the 3-phase winding, as shown in Fig. 5.19.
Each coil is represented as a single phasor and the first coil is set as zero angle. The
angles of other coils are defined with respect to the zero reference by the slot number
in which the coil is laid in. Note that each slot is 2.67° mechanical which equates to
13.33° electrical. For the 4-set coils, the phasors lie between 180° and 270°, but since
the current is reversed in the 4-set coils (due to the coil interconnections), a 180° is
deducted from their angles such that they lie in the first quadrant. In Fig. 5.19, the
4-set coil phasors have lighter color for clarity. Also, note that the amplitude of the
voltages for each coil is normalized to 1.0 p.u.

The pattern is repeated five times, and considering four turns per coil, the per unit
phase voltage for the 3-phase winding is:

$$V_{\mathrm{ph}(3)} = 4 \times 5 \left(V_{A1} + V_{A2} + V_{A3} + V_{A4} + V_{A5} - V'_{A1} - V'_{A2} - V'_{A3} - V'_{A4}\right)$$
$$= 171.98 \angle 26.67°$$

$$(5.15)$$

Employing the same method, then for the 9-phase winding, the phasor diagram for a pattern of 2-coils and 1-coil is shown in Fig. 5.20. The same process explained for the 3-phase winding phasors is used here to obtain the phase voltage. Therefore, the per-phase voltage for the 9-phase winding including 12 turns per coil is:

$$V_{\mathrm{ph}(9)} = 12 \times 5 \left(V_{A1} + V_{A2} - V'_{A1}\right)$$
$$= 179.16 \angle 6.67°$$

$$(5.16)$$

The ratio of 9-phase to 3-phase voltages from the phasor diagram is 1.0417 (179.16/171.98). Hence, the increased voltage in the 9-phase case is due to an improved winding factor. Therefore, an important conclusion here is that compared

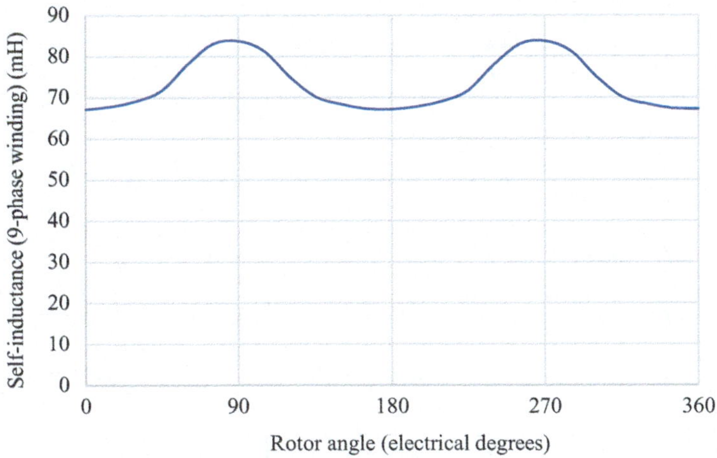

Fig. 5.21 Per-phase self-inductance for one phase of the 9-phase HG WF

Fig. 5.22 Coil representation for 9-phase inductance calculations

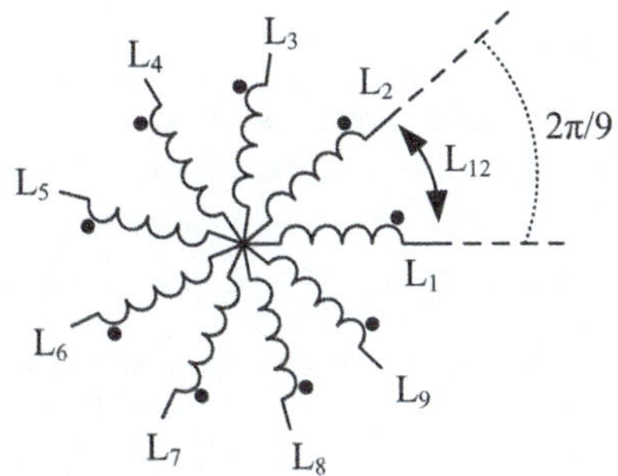

to the 3-phase stator winding for the HG WF, the 9-phase winding results in 4.2% higher output voltage at the machine terminals.

5.4.2 Generator Connected to a Passive Rectifier

Figure 5.21 shows the per-phase self-inductance of the 9-phase winding. The machine inductance changes with rotor angle since the rotor poles are salient. Figure 5.22 shows nine coils each representing one phase where the displacement between adjacent coils is 40 electrical degrees.

Equation 5.17 gives the self- and mutual-inductances for the 9-phase winding. Only the mutual inductance between phase 1 and other phases is shown in Eq. 5.17. Mutual inductances between other phases are obtained similarly.

$$L_1 = L_0^9 + L_g^9 \cos{(2\theta_e)} \qquad M_{12} = L_0^9 \cos{\left(\frac{2\pi}{9}\right)} + L_g^9 \cos{\left(2\theta_e - \frac{2\pi}{9}\right)}$$

$$L_2 = L_0^9 + L_g^9 \cos{\left(2\theta_e + \frac{2\pi}{9}\right)} \qquad M_{13} = L_0^9 \cos{\left(\frac{4\pi}{9}\right)} + L_g^9 \cos{\left(2\theta_e - \frac{4\pi}{9}\right)}$$

$$L_3 = L_0^9 + L_g^9 \cos{\left(2\theta_e + \frac{4\pi}{9}\right)}$$

$$L_4 = L_0^9 + L_g^9 \cos{\left(2\theta_e + \frac{6\pi}{9}\right)} \qquad M_{14} = L_0^9 \cos{\left(\frac{6\pi}{9}\right)} + L_g^9 \cos{\left(2\theta_e - \frac{6\pi}{9}\right)}$$

$$L_5 = L_0^9 + L_g^9 \cos{\left(2\theta_e + \frac{8\pi}{9}\right)} \qquad M_{15} = L_0^9 \cos{\left(\frac{8\pi}{9}\right)} + L_g^9 \cos{\left(2\theta_e - \frac{8\pi}{9}\right)}$$

$$L_6 = L_0^9 + L_g^9 \cos{\left(2\theta_e + \frac{10\pi}{9}\right)} \qquad M_{16} = L_0^9 \cos{\left(\frac{10\pi}{9}\right)} + L_g^9 \cos{\left(2\theta_e - \frac{10\pi}{9}\right)}$$

$$L_7 = L_0^9 + L_g^9 \cos{\left(2\theta_e + \frac{12\pi}{9}\right)} \qquad M_{17} = L_0^9 \cos{\left(\frac{12\pi}{9}\right)} + L_g^9 \cos{\left(2\theta_e - \frac{12\pi}{9}\right)}$$

$$L_8 = L_0^9 + L_g^9 \cos{\left(2\theta_e + \frac{14\pi}{9}\right)} \qquad M_{18} = L_0^9 \cos{\left(\frac{14\pi}{9}\right)} + L_g^9 \cos{\left(2\theta_e - \frac{14\pi}{9}\right)}$$

$$L_9 = L_0^9 + L_g^9 \cos{\left(2\theta_e + \frac{16\pi}{9}\right)} \qquad M_{19} = L_0^9 \cos{\left(\frac{16\pi}{9}\right)} + L_g^9 \cos{\left(2\theta_e - \frac{16\pi}{9}\right)}$$

$$(5.17)$$

where.

L_i: Self-inductance for phase i ($i = 1, \ldots, 9$)
M_{ij}: Mutual inductance between phase i and j ($i = 1, \ldots, 9$ and $j = 1, \ldots, 9$)
L_0^9: Average phase inductance for the 9-phase winding
L_g^9: Phase inductance amplitude variations above average for the 9-phase winding

The phase inductance shown in Fig. 5.21 is unsaturated, and its equivalent synchronous inductance is calculated from Eq. (4.11). However, the saturated synchronous inductance using Eq. (4.13) for the 9-phase HG WF is 112.33 mH.

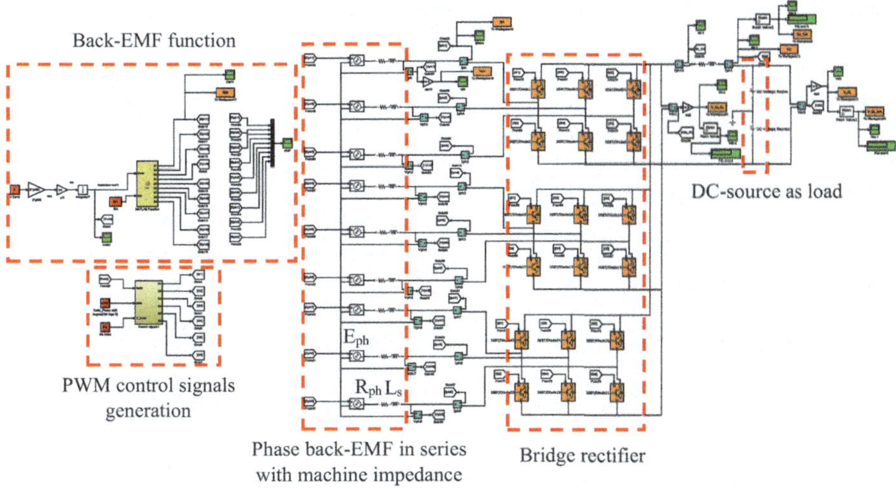

Fig. 5.23 MATLAB model for the 9-phase HG WF connected to a rectifier

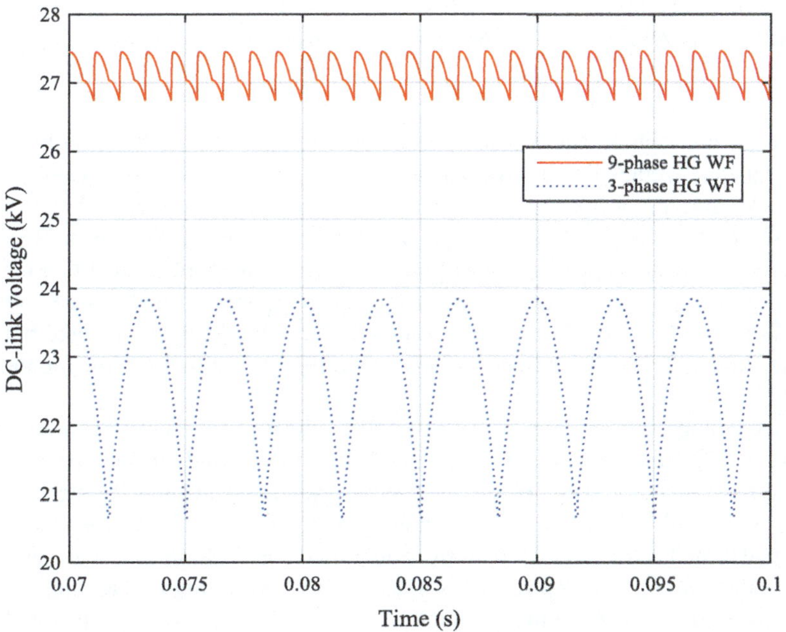

Fig. 5.24 No-load rectified DC-link voltage for the 3- and 9-phase HG WF

The MATLAB Simulink model for the 9-phase HG WF is shown in Fig. 5.23. Note that the gate signals are disabled so that the converter operates as a passive rectifier. The model embeds 3×3-phase converter that are connected to form a 9-phase converter. On the DC link, a battery is used as a load. The no-load rectified DC-link

Table 5.6 Comparison of three-phase and nine-phase machines at an average DC-link voltage of 14 kV

HG WF with	3-phase stator winding	9-phase stator winding	Ratio (9-phase)/ (3-phase)
No-load peak phase voltage (kV)	13.77	13.94	1.0123
No-load RMS voltage (kV)	10.37	10.80	1.0415
RMS phase current (A)	512.23	179.28	0.35
Average phase power (MW)	3.21	1.115	0.35
No-load average DC-link voltage (kV)	22.77	27.32	1.1998
No-load peak DC-link voltage (kV)	23.85	27.45	1.151
No-load minimum DC-link voltage (kV)	20.65	26.74	1.295
No-load peak-to-peak ripples (kV)	3.2 (13.4%)	0.71 (2.6%)	0.22
Average DC current (A)	687.93	715.06	1.04
Peak DC current (A)	721.1	718.7	0.997
Average DC power (MW)	9.63	10.03	1.042
Peak DC power (MW)	10.685	10.06	0.94

voltage for the 9-phase and 3-phase windings of the HG WF is shown in Fig. 5.24. The simulated average DC-link voltage is consistent with the analytical value, that is, 27.32 kV, giving confidence in the model accuracy. The rectified voltage for the 9-phase system is higher than the 3-phase rectified voltage as seen from Fig. 5.24. Moreover, the peak-to-peak voltage ripple is reduced in the 9-phase rectified voltage, hence lower requirements for filtering. Table 5.6 compares 3- and 9-phase voltages. Therefore, the 9-phase HG not only results in 4.2% higher RMS voltage but also 19.98% higher DC-link rectified voltage, therefore a justified option for the HG stator.

Phase voltage, current, and also back-EMF for the 9-phase HG WF connected to a passive rectifier are shown in Fig. 5.25. DC-link power and current are shown in Fig. 5.26. Compared to the 3-phase, the 9-phase system results in 4.2% higher power which is a result of higher phase voltage due to improved winding factor. The important conclusion here is that by going from a 3-phase stator design to a 9-phase design, the HG WF results is 4.2% higher power, higher DC-link voltage, and reduced ripple, hence improvement in power and voltage quality, that is, lower filtering requirement. From machine design point of view, the 9-phase machine is more power dense.

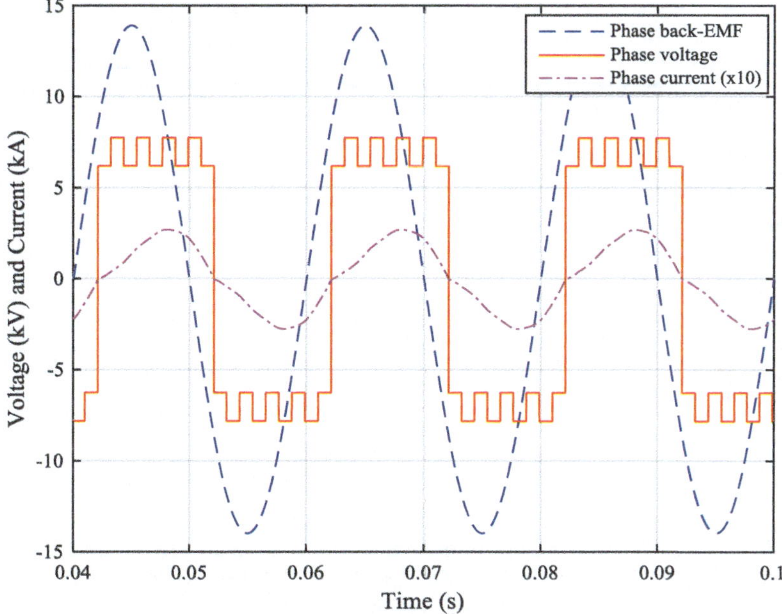

Fig. 5.25 Phase back-EMF, phase voltage, and current for a 9-phase HG WF connected to a passive rectifier and a DC-link voltage of 14 kV

5.5 Design of HG PM Rotor with NdFeB Surface Magnets

Previous design of the HG with 100% WF, that is, HG WF, was discussed. In this section, design of the HG with surface magnet rotor is addressed. The two machines, HG WF and HG with PM rotor, referred to as HG PM, are then combined to form the HG. The HG shares a common stator between WF and PM rotors. Therefore, the 9-phase stator winding designed for the HG WF is used for the HG PM machine. Two different designs for the PM rotor are discussed:

- A surface magnet PM rotor
- An interior magnet PM rotor design

Two different commercially available magnet materials are considered, sintered neodymium–iron–boron (NdFeB) and sintered ferrite. NdFeB is a rare-earth magnet material that has been used in wind applications due to its high-energy density [66, 67]. However, there was a supply issue in 2008 that raised many uncertainties in the source of all rare-earth martials [68, 69], hence the investigation into a sintered ferrite rotor topology.

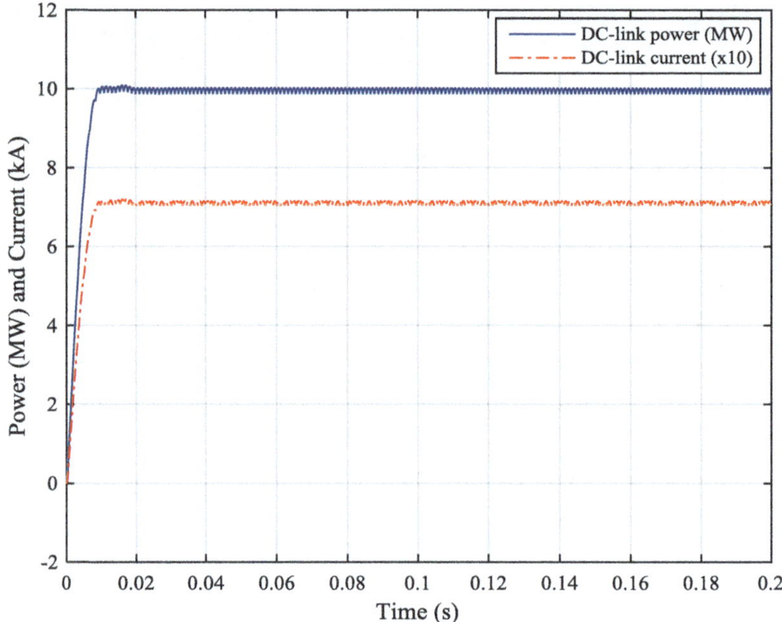

Fig. 5.26 DC-link power and current for a 9-phase HG WF connected to a passive rectifier and with a DC-link voltage of 14 kV

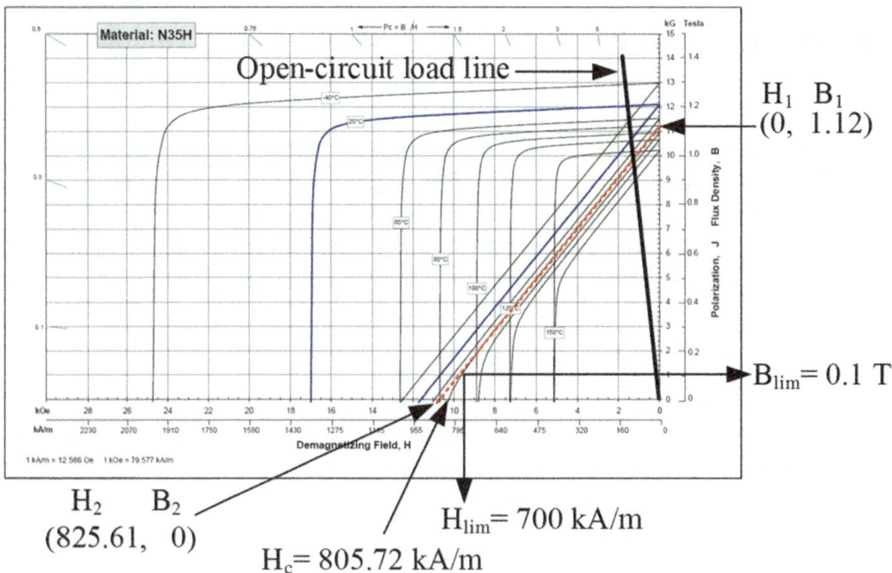

Fig. 5.27 B-H characteristic for NdFeB N35H [70]

Table 5.7 Neodymium–iron–boron magnet N35H properties at 80 °C

Parameter	Symbol	Value
Remanence (T)	B_r	1.12
Relative permeability	μ_r	1.08
Vacuum permeability (H/m)	μ_0	$4\pi \times 10^{-7}$
Coactivity (kA/m)	H_c	805.72
Magnetic intensity demagnetization limit (kA/m)	H_{lim}	700
Flux-density demagnetization limit (T)	B_{lim}	0.1
Magnet electrical conductivity (MS/m)	σ_{m}	0.556

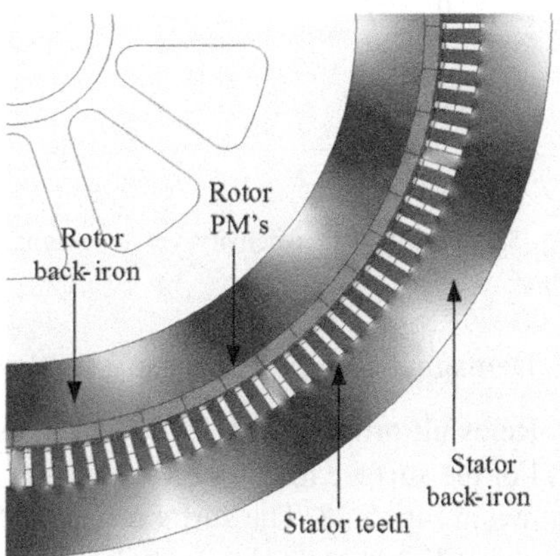

Fig. 5.28 Surface PM rotor structure

5.5.1   Sintered NdFeB Characteristics

The intrinsic and normal characteristics for a typical high-energy sintered NdFeB magnet supplied by Arnold Magnetics UK [70], referred to as N35H, are shown in Fig. 5.27. The magnet characteristics are a function of temperature; hence, the machine internal and external operating ambient are important to understand. The magnet operating point lies on the linear characteristic and its intersection with the open-circuit load line. The slope of the linear B–H characteristic is related to the relative permeability and vacuum permeability, hence at machine operating temperature 80 °C (dotted line in Fig. 5.27):

$$\mu_r = \left(\frac{1}{\mu_0}\right)\left(\frac{B_2 - B_1}{H_2 - H_1}\right) \tag{5.18}$$

where

μ_r: Magnet relative permeability
μ_0: Vacuum permeability (H/m)
B: Flux-density (T)

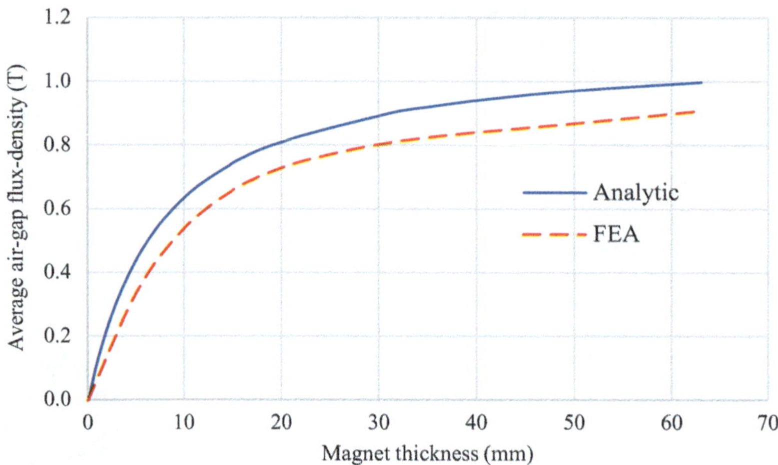

Fig. 5.29 Air-gap flux-density versus magnet thickness for sintered NdFeB surface magnet PM rotor

H: Demagnetizing field intensity (kA/m)

Relevant properties of N35H are detailed in Table 5.7.

For the surface magnet PM rotor, the magnets are mounted on the rotor surface as shown in Fig. 5.28. The surface magnet PM machine is designed to have 10 poles, as for the WF, but the poles are full pitched, that is, cylindrical rotor as opposed to the salient WF rotor. Due to the circumferentially long magnet length (670.968 mm), each magnet is split into six equal sections with each section magnetized separately. This will prevent the magnet being magnetized unevenly which would result in lower flux-density in some locations in the magnet pole.

For this large diameter machine, the magnet and air-gap surface areas can be assumed equal. Hence, the magnet flux-density can be estimated from:

$$B_m = \frac{B_r}{1 + \mu_r \frac{l_g}{l_m}} \ (\text{T}) \tag{5.19}$$

By rearranging Eq. (5.19), the magnet thickness is:

$$l_m = \frac{\mu_r l_g}{\frac{B_r}{B_m} - 1} \ (\text{mm}) \tag{5.20}$$

where

l_g: Air-gap radial thickness (mm)

Table 5.8 Magnet thickness and average air-gap flux-density for NdFeB surface magnet PM rotor

Magnet thickness, l_m (mm)	Average air-gap flux-density, B_m (T)	
	Analytic	FEA
25	0.86	0.772
30	0.9	0.804
35	0.92	0.826
40	0.94	0.842

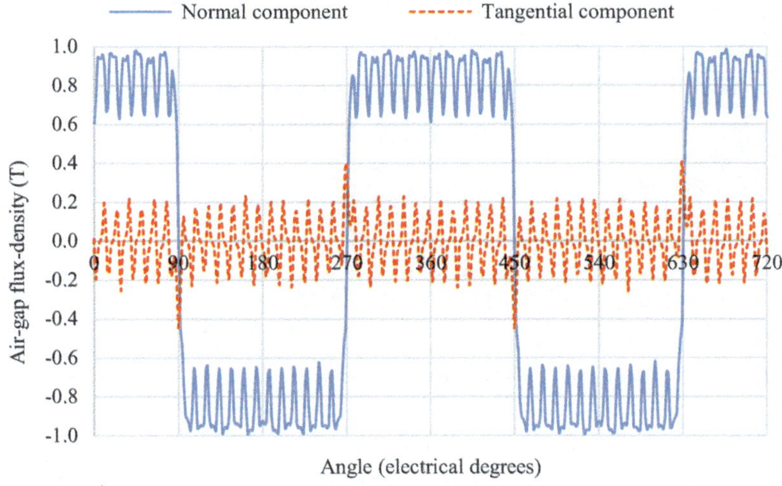

Fig. 5.30 Air-gap flux-density for NdFeB surface magnet PM rotor

The air-gap length is 7 mm, consistent with the minimum air-gap dimension for the WF rotor. Using the values in Table 5.7, the magnet thickness in terms of the magnet flux-density is plotted in Fig. 5.29.

The average air-gap flux-density for the WF rotor was 0.998 T, as previously shown in Table 4.8. For this average air-gap flux-density, the magnet length from Eq. (5.20) would be 61.84 mm. At this flux-density, the stator back-iron and tooth are saturated for the HG WF as seen from Fig. 4.26. While the WF is highly loaded, it is not sensible to run the PM stator to the same load since this is an effective waste of the rotor magnet material. Hence, considering reduced magnet volume, the magnet length is reduced and the stator tooth flux-density reduced to 1.85 T, corresponding to an average air-gap flux-density of 0.854 T, as detailed in Table 4.8 (Chap. 4). The magnet thickness is now analytically calculated to be 25 mm. However, FEA results show that with a magnet thickness of 25 mm, the average air-gap flux-density is 0.77 T which is 10% lower than the targeted 0.854 T. This is due to flux-leakage not accounted for by the simple analytic expression of Eq. (5.19). Therefore, the magnet thickness is increased to obtain the target average flux-density. Table 5.8 lists different magnet thicknesses and the average air-gap

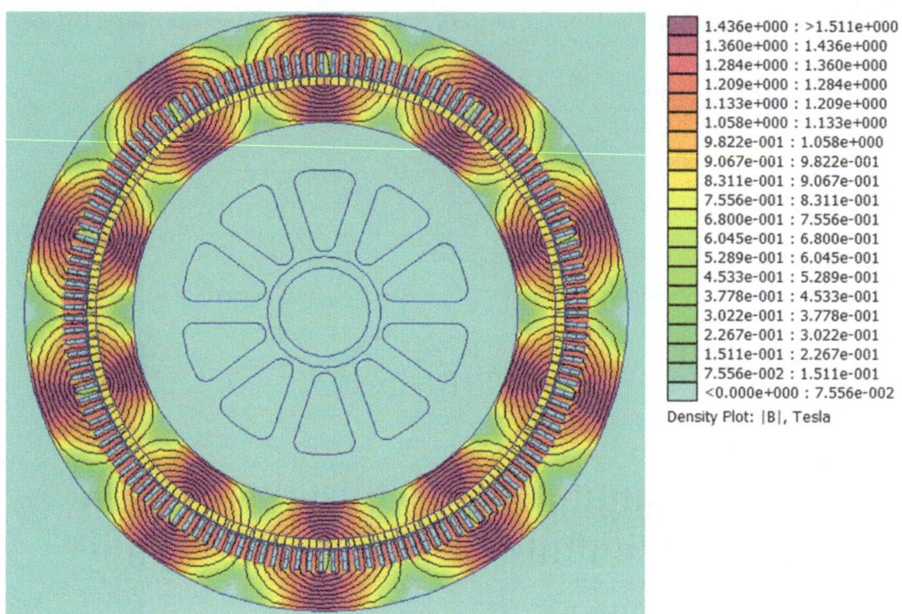

Fig. 5.31 Field distribution for NdFeB surface magnet PM rotor

Table 5.9 Flux-density for different sections for NdFeB surface magnet PM rotor

Section	Flux-density (T)
Average air-gap	0.842
Stator tooth	1.27
Stator back-iron	1.44
Rotor back-iron	1.72

flux-density calculated analytically and from FEA. The magnet thickness is therefore chosen to be 40 mm which produces an average flux-density of 0.842 T in the air gap. Figure 5.30 presents the air-gap flux-density for the surface magnet rotor design for both tangential and normal components. Figure 5.31 illustrates a typical open-circuit field distribution for the machine with surface magnets, while Table 5.9 presents average flux-density in different sections of the machine. The flux-density for the tooth and back-iron is denoted for the sections with the highest density.

From the N35H characteristics, the minimum flux-density for the magnet is 0.1 T as stated in Table 5.7. The minimum magnet thickness in order to prevent magnet demagnetization is calculated as follows:

$$l_m^{\min} = \frac{-l_g}{\mu_0 \, H_{\lim}} - \mu_r \, l_g \;\; (\text{mm}) \tag{5.21}$$

yielding a minimum magnet length of 0.4 mm to prevent demagnetization, a value much below the 40 mm design length. In order to verify this, an FEA file is set up,

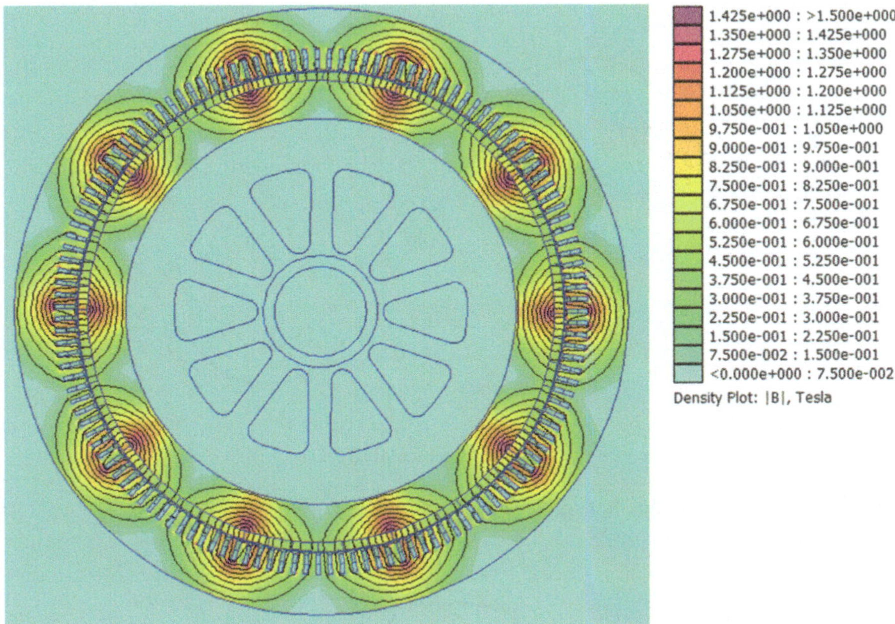

Fig. 5.32 Demagnetizing field distribution for NdFeB surface magnet PM rotor

Table 5.10 Sintered NdFeB magnet dimensions for the surface magnet PM machine

Item	Value (mm)	
Magnet length	670.968	
Magnet radial thickness, l_m	40	
Maximum banding material stress (MPa)	Banding thickness (mm)	
	1.50	2.00
Carbon fiber[a]	1011.20	765.51
Kevlar [a]	1016.30	770.62

[a]The yield stress is taken as 1100 MPa

Fig. 5.32, with full-load stator current in the 9-phase winding (peak phase current $=$ 170.67 A). The PM rotor is rotated such that the magnet field is directly opposed by the stator field at each pole. This condition represents a direct symmetrical short circuit at the machine terminals. After solution, the FEA results file is scrutinized and the weakened magnet flux-density is obtained at different locations inside the magnet. The minimum FEA-measured flux-density in the magnet is 0.35 T, which is higher than the demagnetization limit of 0.1 T. Therefore, the magnets will not be demagnetized under this worst load condition. Note that the seeming high magnet length is not to counter worst-case stator reaction field but to achieve the average air-gap flux-density while maintaining a mechanical air-gap of 7 mm. This air gap may be reduced with the ensuing reduction of magnet length (l_m);

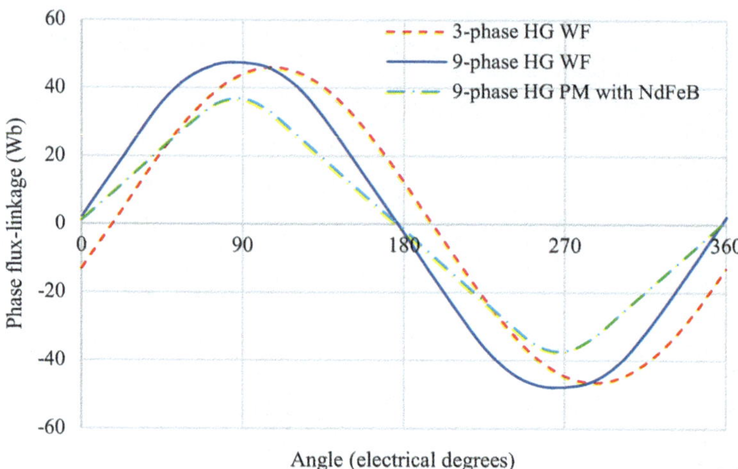

Fig. 5.33 Phase winding flux-linkage versus rotor position

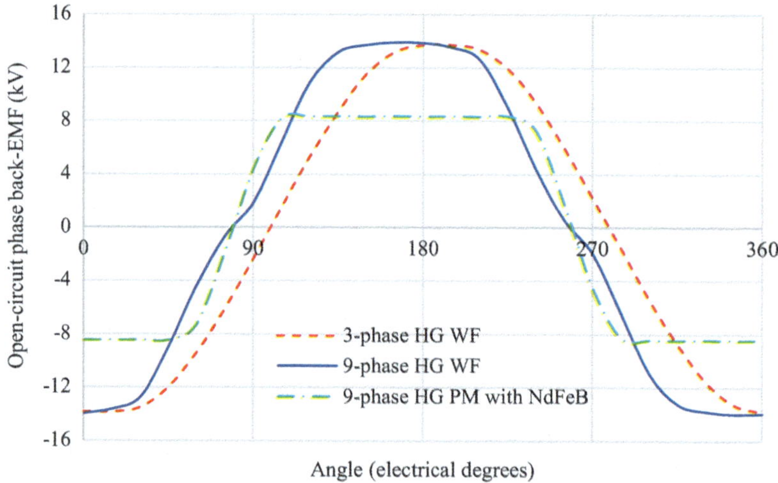

Fig. 5.34 Phase winding open-circuit back-EMFs

however, it was thought prudent to maintain the working mechanical air gap as per
the benchmark SG.

5.5.2 *Flux-Linkage, Back-EMF, and Torque*

The surface magnet PM machine has 10 rotor poles, each split into 6 sections and
magnetized separately in a fixed axis relative to the magnet piece centerline. The
air-gap thickness, stator inner and outer diameter, and stator structure for the HG PM

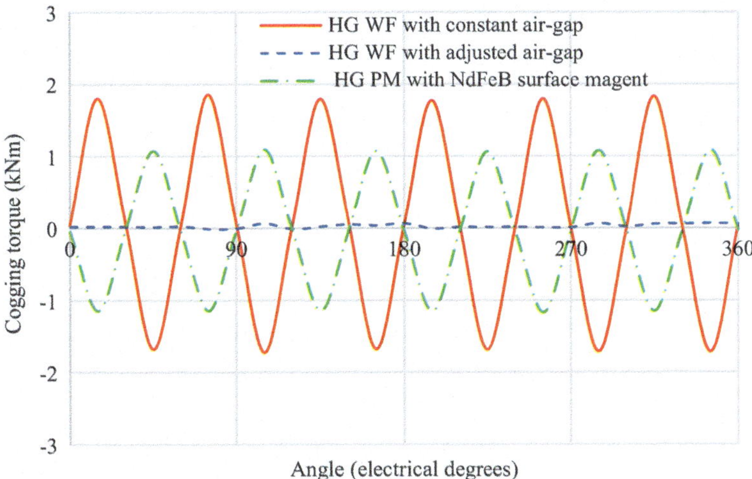

Fig. 5.35 Cogging torque waveforms

are the same as the HG WF. In order to facilitate the comparison, the axial length for the HG PM is chosen to be the same as the HG WF, that is, 400 mm. Later on, these two machines, that is, WF and PM, are combined and scaled accordingly to obtain the final HG design. Table 5.10 lists magnet length and thickness for the surface magnet PM machine. In order to protect the magnets from damages due to centrifugal forces, the periphery of the surface PM rotor is covered with a layer of banding applied under tensile of stress. Table 5.10 shows maximum stress in the banding for two different banding thicknesses and two typical banding materials, carbon fiber and Kevlar. The peripheral velocity of the rotor at the nominal rotor speed of 600 RPM is 67.5 m/s. A 20% overspeed, that is, 720 RPM, is considered to give a margin for safety.

FEA results for the winding per-phase flux-linkage for the HG PM with NdFeB surface magnets are presented in Fig. 5.33, while Fig. 5.34 shows the phase open-circuit back-EMF at a nominal rotor speed of 600 RPM. For the purpose of comparison, waveforms for the 3-phase HG WF and 9-ph HG WF are also shown. The difference in shape of the 9-phase HG PM and HG WF back-EMF arises from the different rotor structures, that is, cylindrical rotor with full-pitched magnets for the HG PM compared to salient pole rotor for the HG WF. The peak back-EMF for 9-phase HG PM is lower compared to the 9-phase HG WF because the HG PM operates at lower air-gap flux-density (0.842 T) compared to 0.998 T for the HG WF.

The cogging torque (at zero stator current) for the HG PM is plotted in Fig. 5.35. For the purpose of comparison, the HG WF cogging torque for both constant and adjusted air-gaps is shown in Fig. 5.35. The peak cogging torque is lower for the HG PM due to lower air-gap flux-density. However, the HG PM is higher than that of the adjusted air-gap HG WF. This cogging torque could be reduced by reducing the rotor pole arc to pole pitch, but since this would not significantly affect the total

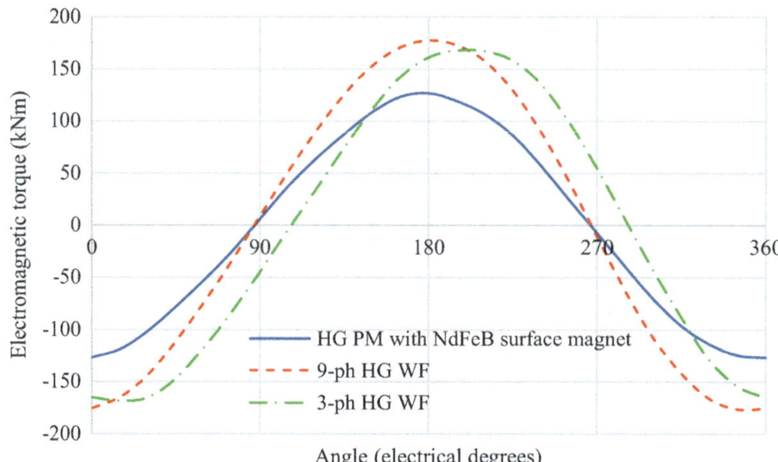

Fig. 5.36 Electromagnetic torque at full-load peak stator current 170.67 A for 9-phase and 512 A for 3-phase, and rotor field current 486.23 A for the HG WF

torque or machine power capability, no further work was carried out here. Needless to say, this would be a future modification to any final machine design.

The FEA results for the machine electromagnetic torque at full-load stator sinusoidal current excitation are presented in Fig. 5.36. The 9-phase full-load peak stator current for the HG PM and 9-phase HG WF is 170.67 A, while it is 512 A for the 3-phase HG WF. The nominal field current for the 3- and 9-phase HG WF is 486.23 A. The peak torque for the 9-phase HG WF is 4.2% higher than that of the 3-phase HG WF due to the improved winding factor as discussed previously. However, the HG PM torque is lower due to lower air-gap flux-density.

5.5.3   Inductances

The HG PM with a surface magnet design has a cylindrical rotor. Therefore, the machine per-phase inductance is essentially constant with respect to rotor position. Indeed, any perturbations due to stator slots are neglected by the large effective magnetic air-gap posed by the NdFeB magnets and mechanical air-gap. In order to calculate the machine inductance, the magnets are set to air, and the rotor is rotated while one phase is excited with full-load stator current. Figure 5.37 shows the per-phase self-inductance and mutual inductances between phase 1 and other phases for the HG PM with surface magnets. It is seen that the HG PM inductances are constant as expected.

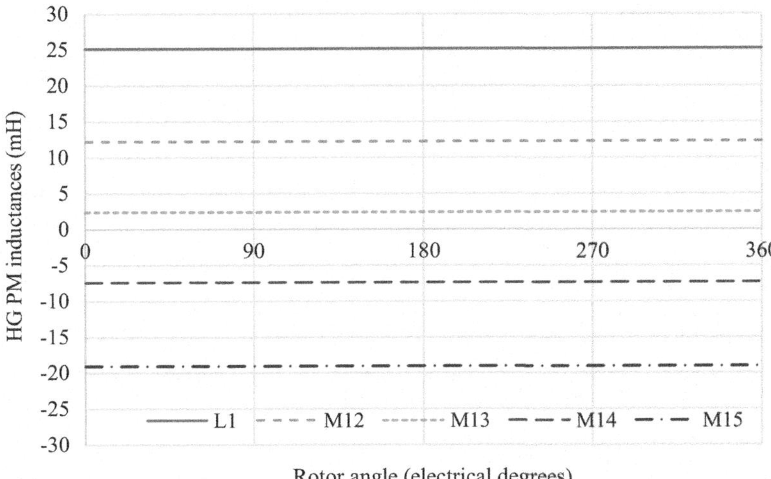

Fig. 5.37 Self- and mutual-inductances for the HG PM with surface magnets

By setting the variable part of inductance to zero in Eq. (5.17), the self- and mutual-inductances for the 9-phase winding are obtained:

$$M_{12}^{\text{PMs}} = L_1^{\text{PMs}} \cos\left(\frac{2\pi}{9}\right) \qquad M_{13}^{\text{PMs}} = L_1^{\text{PMs}} \cos\left(\frac{4\pi}{9}\right)$$

$$\begin{aligned} L_1^{\text{PMs}} = L_2^{\text{PMs}} = L_3^{\text{PMs}} = L_4^{\text{PMs}} \\ = L_5^{\text{PMs}} = L_6^{\text{PMs}} = L_7^{\text{PMs}} \\ = L_8^{\text{PMs}} = L_9^{\text{PMs}} \end{aligned} \quad \begin{aligned} M_{14}^{\text{PMs}} = L_1^{\text{PMs}} \cos\left(\frac{6\pi}{9}\right) \\ M_{16}^{\text{PMs}} = L_1^{\text{PMs}} \cos\left(\frac{10\pi}{9}\right) \end{aligned} \quad \begin{aligned} M_{15}^{\text{PMs}} = L_1^{\text{PMs}} \cos\left(\frac{8\pi}{9}\right) \\ M_{17}^{\text{PMs}} = L_1^{\text{PMs}} \cos\left(\frac{12\pi}{9}\right) \end{aligned}$$

$$M_{18}^{\text{PMs}} = L_1^{\text{PMs}} \cos\left(\frac{14\pi}{9}\right) \qquad M_{19}^{\text{PMs}} = L_1^{\text{PMs}} \cos\left(\frac{16\pi}{9}\right)$$

$$(5.22)$$

Note that:

$$\begin{aligned} M_{12}^{\text{PMs}} = M_{19}^{\text{PMs}} \qquad M_{13}^{\text{PMs}} = M_{18}^{\text{PMs}} \\ M_{14}^{\text{PMs}} = M_{17}^{\text{PMs}} \qquad M_{15}^{\text{PMs}} = M_{16}^{\text{PMs}} \end{aligned}$$

Referring to Fig. 5.22, the mutual inductances for the HG PM with surface magnets can be generalized as follows:

Table 5.11 Inductances for the HG PM with surface magnets

Item	Symbol	Value (mH)
Self-inductance	L_i^{PMs}	25.23
Mutual inductances	$M_{i,i+1}^{\text{PMs}} = M_{i,i+8}^{\text{PMs}}$	12.26
	$M_{i,i+2}^{\text{PMs}} = M_{i,i+7}^{\text{PMs}}$	2.42
	$M_{i,i+3}^{\text{PMs}} = M_{i,i+6}^{\text{PMs}}$	-7.39
	$M_{i,i+4}^{\text{PMs}} = M_{i,i+5}^{\text{PMs}}$	-19.06
Synchronous inductance	L_s^{PMs}	113.35

$$M_{i,i+1}^{\text{PMs}} = M_{i,i+8}^{\text{PMs}} = L_1^{\text{PMs}} \cos\left(\frac{2\pi}{9}\right) \quad M_{i,i+2}^{\text{PMs}} = M_{i,i+7}^{\text{PMs}} = L_1^{\text{PMs}} \cos\left(\frac{4\pi}{9}\right)$$

$$M_{i,i+3}^{\text{PMs}} = M_{i,i+6}^{\text{PMs}} = L_1^{\text{PMs}} \cos\left(\frac{6\pi}{9}\right) \quad M_{i,i+4}^{\text{PMs}} = M_{i,i+5}^{\text{PMs}} = L_1^{\text{PMs}} \cos\left(\frac{8\pi}{9}\right) \quad M_{ij}^{\text{PMs}}$$

$$= M_{ji}^{\text{PMs}} \tag{5.23}$$

where

L_i^{PMs}: Self-inductance for i^{th} phase for the HG PM with surface magnets

M_{ij}^{PMs}: Mutual inductance between phases i and j for the HG PM with surface magnets

The synchronous inductance for the HG PM with surface magnet is calculated from Eq. (4.11). The HG PM is designed such that operates in the linear magnetic region; hence, the unsaturated inductance is only considered in the machine analysis. Table 5.11 lists inductances for the HG PM with surface magnets.

5.6 Design of HG PM Rotor with Ferrite-Embedded Magnets

In this section, design of an HG PM machine with embedded magnet rotor using sintered ferrite magnets is addressed. The B–H characteristics for a typical sintered ferrite magnet manufactured by Arnold Magnetics, referred to as Ferrite AC-5, are shown in Fig. 5.38. The dotted line shows the desired characteristics at a nominal machine operating temperature of 80 °C. Properties of the ferrite magnet at 80 °C are detailed in Table 5.12. Ferrite relative permeability is calculated using two points on the graph in Fig. 5.38 and Eq. (5.18).

The sintered ferrite magnet remanence at 80 °C (0.33 T) is much lower than that of the sintered NdFeB magnet (1.12 T). Therefore, if the ferrite magnets are to replace the surface magnets in the HG PM design, the air-gap flux-density will be lower than that of HG PM with NdFeB surface magnets by a factor close to the ratio of their respective remanences. This would require a longer machine axial length in

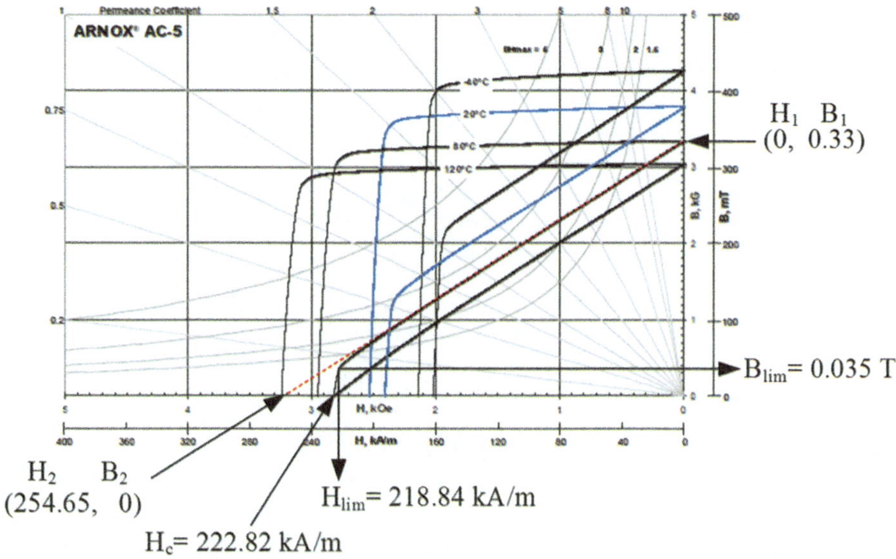

Fig. 5.38   B–H characteristic for Ferrite AC-5 [70]

Table 5.12   Ferrite magnet AC-5 properties at 80 °C

Parameter	Value
Remanence (T)	0.33
Relative permeability	1.03
Coactivity (kA/m)	222.82
Magnetic intensity demagnetization limit (kA/m)	218.84
Flux-density demagnetization limit (T)	0.035
Magnet electrical conductivity (MS/m)	0

order to obtain the same back-EMF and therefore not an ideal option for the HG PM design. However, the HG PM with embedded ferrite magnets in a "high-field" topology is studied here to investigate the feasibility of such design. Figure 5.39 shows an HG PM with ferrite "high-field" embedded rotor design.

The ferrite magnets are arranged to realize a 10-pole rotor field. Basic analytic sizing is used to estimate the initial thickness and length of the ferrite magnets for the embedded PM rotor design. The outer diameter of the rotor for the embedded magnet design is the same as surface magnet rotor, but the inner rotor diameter has to be larger to accommodate the embedded magnets; hence, the amount of rotor material is greater compared to that of surface magnet rotor. There are 10 poles; hence, 10 ferrite magnets displaced 36 mechanical degrees from one another. Each ferrite magnet is magnetized circumferentially as indicated in Fig. 5.40. The air-gap thickness is maintained at 7 mm, as per the other machine designs. The no-load field distribution for the HG PM with an embedded magnet rotor is shown in Fig. 5.40.

Ideally, it is expected that fluxes due to two adjacent ferrites add up in the air-gap to increase the air-gap flux-density. However, for the same magnet thickness and

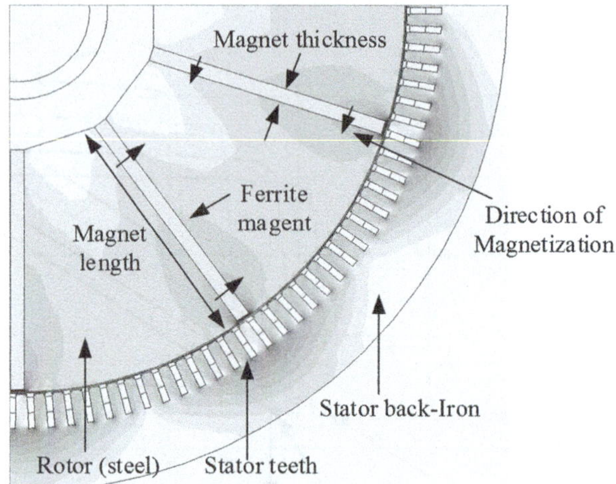

Fig. 5.39 Ferrite magnet arrangement in HG PM "high-field" embedded rotor design

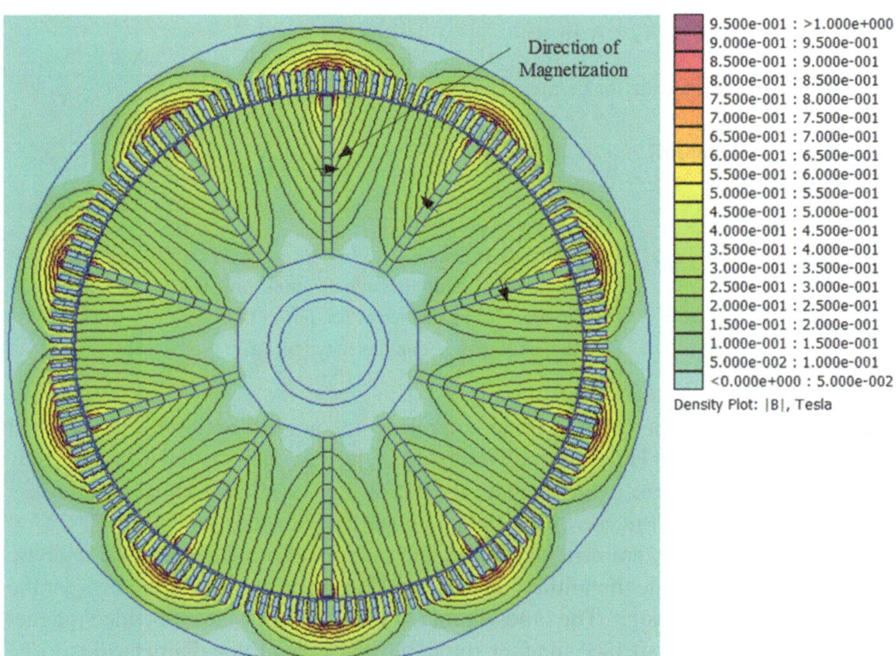

Fig. 5.40 Typical no-load magnetic field distribution for the HG PM with embedded sintered ferrite magnets

length as in NdFeB magnets, the average air-gap flux-density is 0.274 T. As seen from Fig. 5.40, the fluxes don't add up. Therefore, in order to increase the air-gap flux-density, the ferrite magnet's thickness is increased and average air-gap flux-density is calculated using FEA. Table 5.13 shows average air-gap flux-density for

Table 5.13 Magnet size and average air-gap flux-density for the HG PM with embedded sintered ferrite PM rotor

Magnet thickness (mm)	Average air-gap flux-density (T)	
	Magnet length = 670.968 mm (full length)	Magnet length = 335.484 mm (half length)
40	0.274	0.163
60	0.324	0.183
80	0.357	0.196
100	0.38	0.204
120	0.4	0.211
140	0.411	0.216
160	0.422	0.220

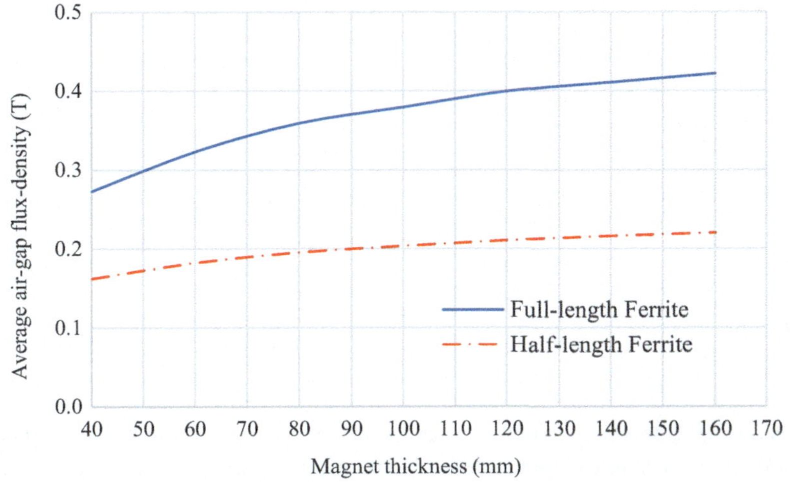

Fig. 5.41 HG PM average air-gap flux-density with respect to embedded ferrite magnet size

different ferrite thickness and for two different lengths, (i) a full magnet length of 670.968 mm and (ii) a half magnet length of 335.484 mm, while Fig. 5.41 plots these results graphically.

From Fig. 5.41, it is evident that the air-gap flux-density does not increase linearly with the ferrite magnet thickness. Indeed, it is concluded that the geometry of the rotor is such that the flux due to adjacent ferrites does not add up. In order to investigate the effect of geometry on the air-gap flux-density, the number of poles are increased. Table 5.14 shows average air-gap flux-density for different pole numbers, and Fig. 5.42 illustrates this graphically. As seen, when the number of poles is increased, the air-gap flux-density is increased. Note that the additional poles have the same shape, length (670.968 mm), and thickness (40 mm), and when adding new poles, the angular distance between poles is calculated accordingly to accommodate appropriate amount of ferrite magnets in the rotor constraint. Thus, for the machine

Table 5.14 Average air-gap flux-density for the HG PM with embedded PM rotor and different number of poles (ferrite length 670.968 mm and thickness 40 mm)

Number of poles	Average air-gap flux-density (T)
10	0.274
12	0.299
14	0.32
16	0.337
18	0.351
20	0.361

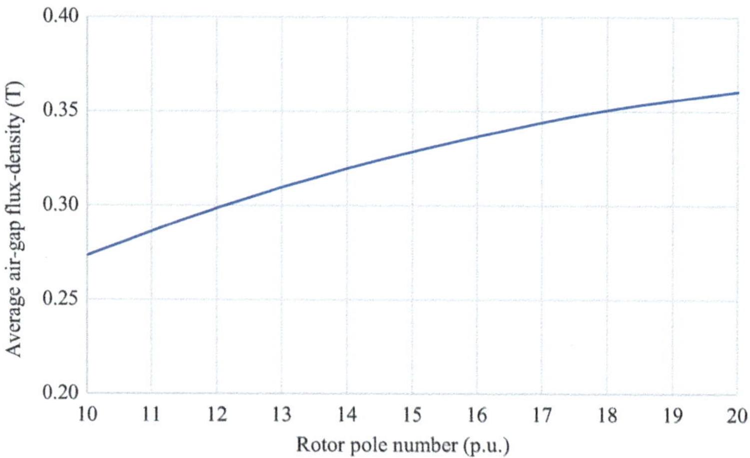

Fig. 5.42 HG PM average air-gap flux-density with respect to number of embedded ferrite magnet rotor poles

dimensions fixed by the benchmark machine design, a 10-pole solution would not be a suitable one for employing a sintered ferrite rotor design.

Therefore, it is concluded that the embedded magnet rotor design with sintered ferrite magnets for the HG PM rotor is not an ideal option and hence will not be investigated further.

5.7   Design of HG PM Rotor with NdFeB-Embedded Magnets

An HG PM rotor design with sintered NdFeB surface magnets was discussed previously. It was also concluded that the embedded rotor design with sintered ferrite magnets is not a suitable option for the HG PM. In this section, an HG PM rotor design with embedded sintered NdFeB magnets is discussed to give some assessment for this rotor topology. The same NdFeB magnet grade as in surface rotor design, that is, N35H, manufactured by Arnold is chosen. The magnets are laid out within the rotor assembly as with the ferrites design. The magnet length

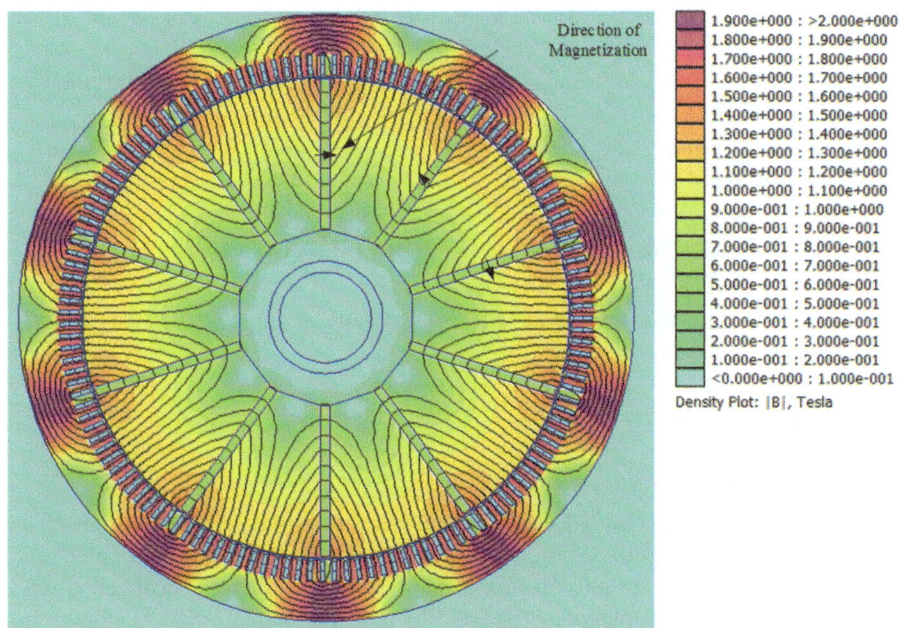

Fig. 5.43 No-load flux distribution for the HG PM with embedded NdFeB rotor design

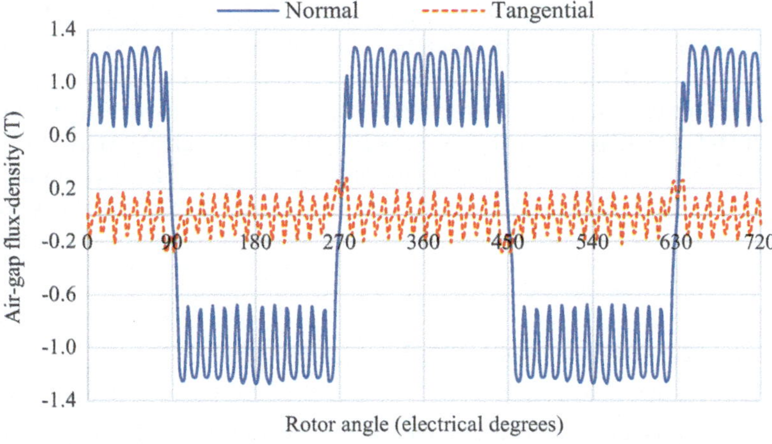

Fig. 5.44 No-load air-gap flux-density with respect to rotor electrical angle for the HG PM with embedded NdFeB rotor design

(670.968 mm) and thickness (40 mm) are as for the surface magnet design. Figure 5.43 shows the no-load flux distribution for the HG PM with embedded NdFeB rotor design, while Fig. 5.44 presents machine air-gap flux-density for two electrical cycles.

Table 5.15 Average air-gap flux-density for the HG PM with different rotor designs (for all designs magnet length is 670.968 mm and thickness 40 mm)

HG PM with	Average air-gap flux-density (T)
Surface NdFeB magnets	0.842
Embedded ferrite magnets	0.274
Embedded NdFeB magnets	1.00

Table 5.16 Flux-density for different sections for the HG PM with NdFeB magnets

Section	Flux-density (T)		
	HG PM with surface NdFeB	HG PM with embedded NdFeB	HG WF
Average air-gap	0.842	1.00	0.998
Stator tooth	1.27	1.54	2
Stator back-iron	1.44	1.81	1.8
Rotor back-iron	1.72	1.12	1.5

Comparing Fig. 5.44 with Fig. 5.30, it is seen that the air-gap flux-density is higher for the embedded design with the same magnet length and thickness. Table 5.15 shows average air-gap flux-density for three PM rotor designs: (i) HG PM with surface NdFeB magnets, (ii) HG PM with embedded ferrite magnets, and (iii) HG PM with embedded NdFeB magnets. Average air-gap flux-density for the embedded NdFeB rotor design is 1.0 T, that is, 18.8% higher than that of surface magnet design and hence a suitable option for the HG PM rotor design.

Table 5.16 lists average flux-density for different sections of the HG PM machine with embedded NdFeB rotor. The tooth flux-density is measured in the middle of the tooth and rotor and stator back-iron flux-densities are for the areas with highest saturation. Note that the average air-gap and stator back-iron flux-density for the HG PM with embedded NdFeB is the same as HG WF. However, the stator tooth and rotor back-iron flux-density is higher in the HG WF machine due to the higher-field loading. The RMS air-gap flux-density for the HG PM with embedded NdFeB magnets is 1.04 T which is 11% lower compared to 1.17 T for the HG WF machine.

5.7.1 Flux-Linkage, Back-EMF, and Torque

The HG PM with embedded NdFeB magnets has the same machine stator as the HG PM with surface magnets, that is, a 9-phase design with distributed windings. The air-gap thickness and stator inner and outer diameter are also the same as the HG PM with surface magnets. The outer diameter of the rotor for the embedded magnet design is the same as surface magnet rotor, but the inner rotor diameter is larger for the embedded rotor. For comparison purposes, the machine axial length is the same as HG PM with surface magnets, that is, 400 mm. Figure 5.45 shows per-phase flux-linkage, while the open-circuit back-EMFs at the nominal rotor speed of 600 RPM is presented in Fig. 5.46. The peak phase back-EMF for the HG PM with embedded

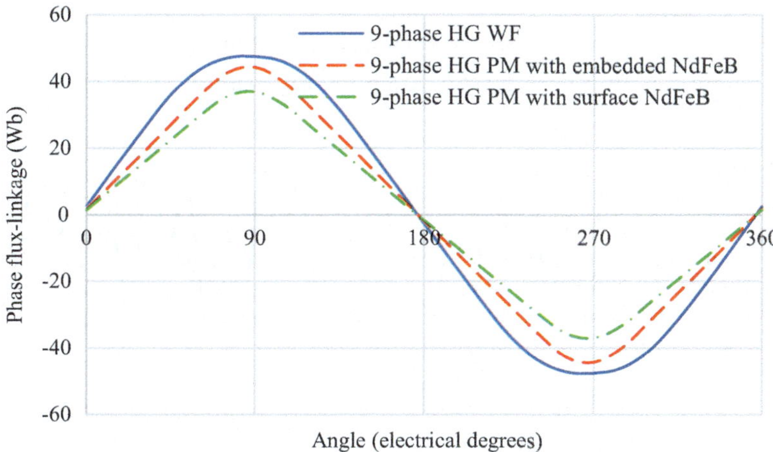

Fig. 5.45 Per-phase flux-linkage for the 9-phase HG WF, 9-phase HG PM with surface NdFeB magnets, and 9-phase HG PM with embedded NdFeB magnets

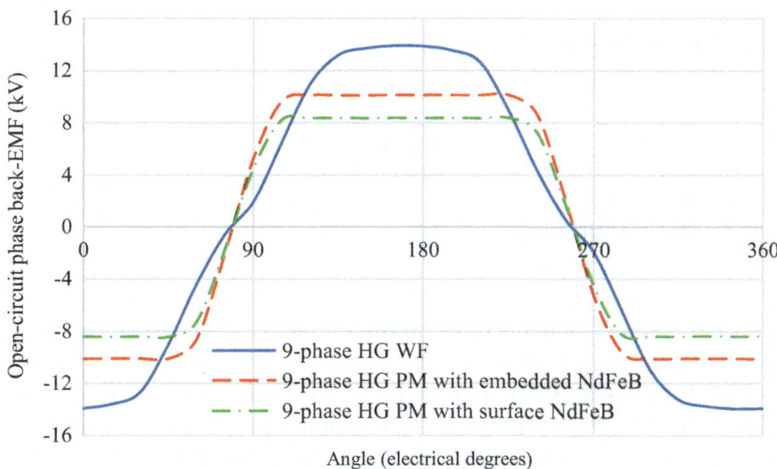

Fig. 5.46 Per-phase open-circuit back-EMF for the 9-phase HG WF, 9-phase HG PM with surface NdFeB magnets, and 9-phase HG PM with embedded NdFeB magnets

Table 5.17 RMS and peak voltages for nine-phase HG with various rotors

	Voltage (kV)	
9-phase stator machine	Peak	RMS
HG WF	13.94	10.80
HG PM with surface NdFeB magnets	8.40	7.70
HG PM with embedded NdFeB magnets	10.13	9.28

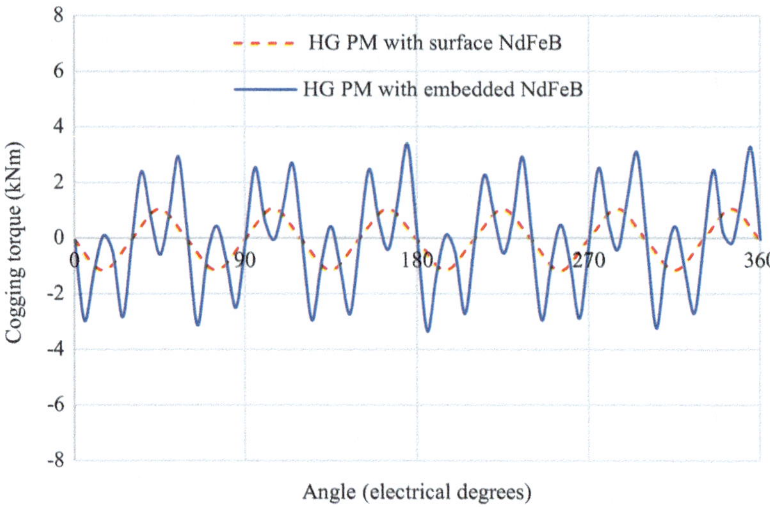

Fig. 5.47 Cogging torque (at zero stator current)

Table 5.18 Peak cogging and total torque for different designs

9-phase stator machine	Peak torque (kNm)		Torque ratio (%) (Cogging/total)
	Cogging	Total	
HG WF	0.08	177.90	0.05
HG PM with surface NdFeB magnets	1.11	127.05	0.9
HG PM with embedded NdFeB magnets	3.40	130.93	2.6

NdFeB is 20.6% higher than that of HG PM with surface magnets. However, the HG PM has lower peak back-EMF compared to the HG WF due to lower flux-linkage. Table 5.17 lists peak and RMS voltages for the HG WP and HG WF.

The machine cogging torque at no-load stator current is plotted in Fig. 5.47. Compared to the surface magnet design, the cogging torque in the embedded rotor has more harmonics with larger amplitudes. Table 5.18 shows peak cogging torque for HG PM with surface and embedded magnets and also for the HG WF. For the HG PM with embedded NdFeB magnets, the peak cogging torque is 2.6% of the machine total torque, whereas the cogging torque in the HG PM with surface magnets is only 0.9% of its total torque. However, the adjusted air-gap HG WF cogging torque is much smaller, that is, 0.05% of the total torque.

The FEA results for machine electromagnetic torque at full-load stator current are presented in Fig. 5.48. The embedded magnet rotor design is electromagnetically a salient pole rotor and hence has two torque components: salient torque and excitation torque. The excitation torque is due to interactions between magnet and stator fields, while the salient torque depends on the machine geometry. The HG PM with surface magnets has a cylindrical rotor; hence, machine torque is only due to excitation. The

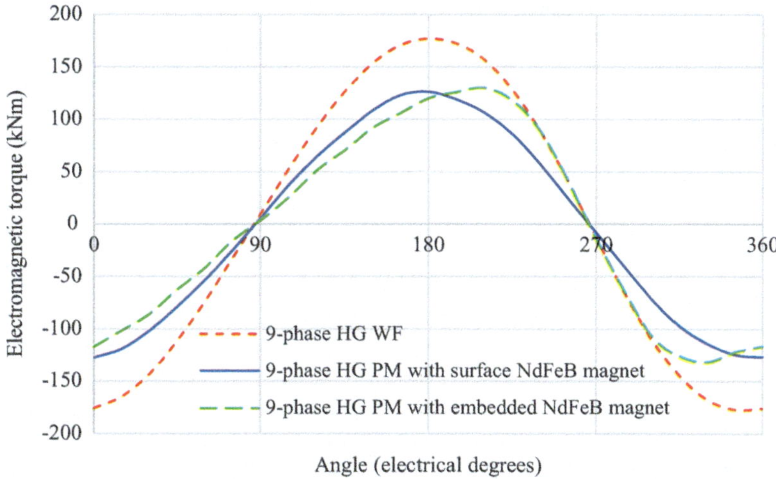

Fig. 5.48 Electromagnetic torque at full-load peak stator current of 170.67 A for the 9-phase stator winding

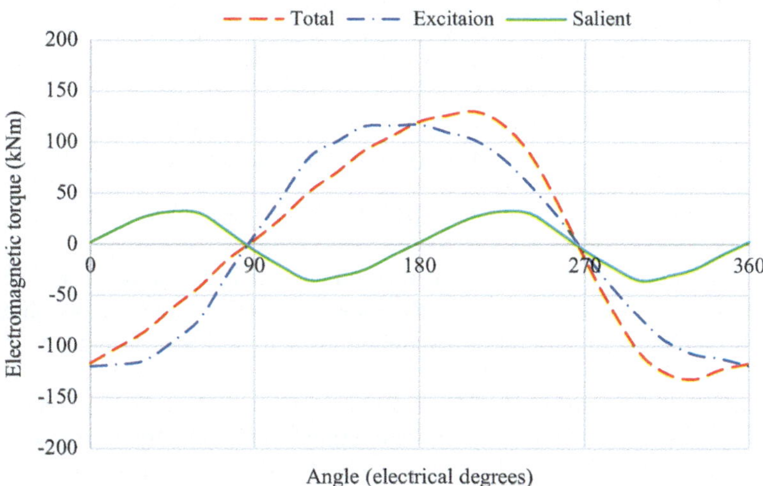

Fig. 5.49 Salient and excitation torque components in the 9-phase HG PM with embedded NdFeB

HG PM with embedded PM has higher torque compared to surface magnet design due to increased air-gap flux-density and saliency component of torque. The HG WF has a salient rotor; however, the machine is highly saturated, and the excitation torque dominates the shape of the torque waveform in Fig. 5.48. In order to estimate the salient torque for the HG PM with embedded NdFeB rotor, the magnets are set to air and the FEA results are plotted in Fig. 5.49. By setting the magnets to air, the machine operating point on the B-H curve changes; hence, the steel operates at a lower flux-density. Therefore, the salient torque calculated is an estimate of the

Table 5.19 Peak salient, excitation, and total torque for different designs

9-phase stator machine	Peak torque (kNm)			Torque ratio (%) (Salient/total)
	Salient	Excitation	Total	
HG WF	52.38	206.40	177.90	29.44
HG PM with surface NdFeB magnets	–	127.05	127.05	0
HG PM with embedded NdFeB magnets	33.41	118.02	130.93	25.52

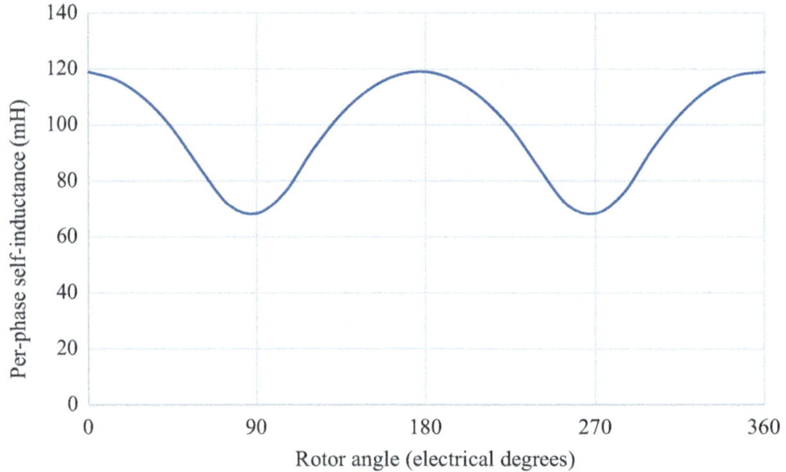

Fig. 5.50 Per-phase self-inductance for the HG PM with embedded NdFeB magnets

actual machine salient torque. By subtracting the salient torque from the total machine torque, the estimation for excitation torque is obtained, as shown in Fig. 5.48. Table 5.19 details peak excitation, salient, and total torque for the HG PM and HG WF.

The peak salient torque is 25.52% of the total peak torque in the HG PM machine with embedded magnets. For the purpose of comparison, the torque components for HG WF and HG PM with surface magnets are also shown in Table 5.19.

5.7.2   Inductances

The HG PM with embedded NdFeB magnets rotor design has a magnetically salient rotor. Therefore, the machine per-phase inductance varies with respect to rotor position. In order to calculate the machine inductance, the magnets are set to air and the rotor is rotated while one phase is excited with full-load stator current. Figure 5.50 shows the per-phase machine inductance. The HG PM with embedded NdFeB self-inductance with respect to rotor electrical angle can be expressed as:

Table 5.20 HG PM with embedded NdFeB inductances

Parameter	Value (mH)
L_0^{PMe}	93.68
L_2^{PMe}	25.31
L_d^{PMe}	102.56
L_q^{PMe}	178.49
L_s^{PMe}	421.56

$$L_{\mathrm{self}}^{\mathrm{PMe}} = L_0^{\mathrm{PMe}} + L_2^{\mathrm{PMe}} \cos\left(2\theta_e\right) \tag{5.24}$$

where

L_0^{PMe}: Average phase inductance

L_2^{PMe}: Phase inductance amplitude variations above average

Compared to the HG WF self-inductance, the d–q axis is switched in the HG PM with embedded NdFeB. Therefore:

$$
\begin{aligned}
L_q^{\mathrm{PMe}} &= \frac{3}{2}\left(L_0^{\mathrm{PMe}} + L_2^{\mathrm{PMe}}\right) \\
L_d^{\mathrm{PMe}} &= \frac{3}{2}\left(L_0^{\mathrm{PMe}} - L_2^{\mathrm{PMe}}\right)
\end{aligned}
\tag{5.25}
$$

The machine equivalent synchronous inductance is calculated from Eq. (4.11). Table 5.20 lists different inductances for the HG PM with embedded NdFeB.

5.8   Final HG Considerations

The three-phase benchmark SG is rated at 5.52 MW. Therefore, the nominal power of the three-phase HG WF, which is essentially the same machine as the benchmark SG, is 5.52 MW. However, as discussed previously when the machine is rewound with a nine-phase stator, the output voltage increases by 4.2%; hence, the nine-phase HG WF nominal power is 5.75 MW. As shown in Table 5.17, the RMS phase voltage of the nine-phase HG WF is 10.80 kV, higher than that of HG PM with surface and embedded magnets, that is, 7.70 kV and 9.28 kV, respectively, due to their reduced air-gap flux-densities relative to the benchmark machine. All three machines, HG WF, HG PM with surface NdFeB, and HG PM with embedded NdFeB, have the same active axial length, that is, 400 mm, and the same number of turns per coil, that is, 12. Therefore, for the same stator current, the machine output power is linearly changed with the voltage. The output power of the HG PM with surface magnet is reduced by a factor of 0.71 (7.70/10.80), while the output power of the HG PM with embedded magnets is reduced by a factor of 0.86 (9.28/10.80). In order to facilitate the comparison, the three machines are scaled to the

Table 5.21 Main machine specifications at 5.75 MW

Parameter	9-phase machine		
	HG WF	HG PM with surface magnets	HG PM with embedded magnets
Output power (MW)	5.75		
RMS phase voltage (kV)	10.80		
Number of turns per coil (p.u.)	12		
Coil cross section (mm^2)	380.64		
Turn cross section (mm^2)	31.72		
Active axial length (mm)	400	560	464

same voltage and power, that is, 10.80 kV and 5.75 MW. The machine voltage is a function of axial length and number of turns in each coil, that is:

$$\text{EMF} \propto N \, l_a \tag{5.26}$$

where

N: Number of turns per coil (p.u.)
l_a: Active axial length (mm)

Therefore, to obtain an RMS phase voltage of 10.80 kV while keeping the same number of turns, the initial axial length of 400 mm for the HG PM with surface magnets is increased to 560 mm (by a factor 1.4) while the axial length of the HG PM with embedded magnets is increased to 464 mm (by a factor of 1.16). Since the stator current is maintained the same, the machine output power is linearly increased for both HG PM with surface and embedded magnets to 5.75 MW. Table 5.21 summarizes the main specification of the three machines at 5.75 MW.

The HG in the wind generation scheme in Chap. 3 is rated at 3.6 MW. Therefore, the 9-phase 5.75 MW machines need to be scaled to deliver a nominal power of 3.6 MW. The machine torque can be expressed as:

$$T = \frac{\pi \, D^2}{2} \, l_a \, Q \, B_{\text{ave}} \tag{5.27}$$

where

D: Air-gap diameter (mm)
l_a: Active axial length (mm)
Q: Electric loading (A/mm)
B_{ave}: Air-gap average flux-density (magnetic loading) (T)

The machine power is related to its torque via rotational speed. Keeping the same air-gap diameter, electric loading, and air-gap average flux-density, the machine axial length is adjusted to obtain desired output power. However, by changing the axial length, the machine output voltage changes. In order to keep the same output voltage while scaling the machine power, the number of turns needs to be changed. The 5.75 MW 9-phase HG WF axial length is 400 mm; hence, the axial length for a 3.6 MW rated power is 250.43 mm, namely:

$$l_a = \frac{3.6}{5.75} \, (400) = 250.43 \text{ mm}$$

The number of turns per coil for the 5.75 MW HG WF is 12; therefore, the adjusted number of turns for the 3.6 MW machine and for the same output voltage is 19.2, namely:

$$N = \frac{400}{250.43} \, (12) = 19.2$$

However, the number of turns needs to be an integer, that is, 19. This requires the 3.6 MW HG WF axial length to be tuned to 253 mm to obtain the same output voltage. Note that the total coil cross section is kept the same; hence, increasing the number of turn reduces the cross section per turn. Similarly with 19 turns per coil, the axial length for the HG PM with surface NdFeB and the HG PM with embedded NdFeB is 354.3 mm and 293.6 mm, respectively. Table 5.22 summarizes the 3.6 MW machine's axial length, number of turns, coil, and turn cross section.

5.8.1 WF and PM Split Ratio

As discussed, the HG has two rotor sections: WF and PM rotors. The HG output voltage is 75% due to the PM and 25% due to WF sections as discussed previously. Therefore, the separate machines analyzed as HG WF and HG PM need to be combined with appropriate ratios to obtain the HG output voltage at 3.6 MW rated power. The machine RMS output voltage is 10.80 kV. Therefore, the PM rotor section provides 75% of this voltage, that is, 8.1 kV, while the WF rotor section provides the 25%, that is, 2.7 kV. The machine axial length is linearly related to its output voltage as expressed in Eq. (5.26). Therefore, the WF rotor section axial length is 25% of the HG WF axial length, that is, 63.25 mm (0.25 × 253). The axial length of the PM rotor section is 75% of the HG PM, that is, 265.73 mm (0.75 × 354.3), providing the rotor is the PM with surface magnet NdFeB. If the rotor is PM with embedded NdFeB magnets, the axial length is 220.2 mm (0.75 × 293.6). Therefore, two different designs for the final HG are considered:

- Design 1: an HG with surface NdFeB PM rotor
- Design 2: an HG with embedded NdFeB PM rotor

Table 5.22 Main machine specifications at 3.6 MW

Parameter	9-phase machine		
	HG WF	HG PM with surface magnets	HG PM with embedded magnets
Output power (MW)	3.6		
RMS phase voltage (kV)	10.80		
Number of turns per coil (p.u.)	19		
Coil cross section (mm^2)	380.64		
Turn cross section (mm^2)	20.03		
Active axial length (mm)	253	354.3	293.6
Mass (kg)[a]			
Stator steel (back-iron + teeth)	3536.14	4952	4103.6
Rotor pole steel	1912.68	–	–
Rotor back-iron steel	2182.82	2701.73	6406.94
Rotor pole PM[b]	–	699.58	589.78
Stator copper	1036	1129.23	1073.41
Rotor copper	774.14	–	–
Total steel	7631.64	7653.73	10510.54
Total copper	1810.14	1129.23	1073.41
Exciter	1141.25	–	–
Total[c] (tonnes)	10.58	9.48	12.18

[a]Steel (lamination) density $= 7650$ kg/m^3; copper density $= 8960$ kg/m^3
[b]PM mass density 7500 kg/m^3
[c]Excluding shaft, machine casing, end caps, cooling system, and other support structure

The WF is the same for both designs. The total HG axial length for HG with surface NdFeB PM rotor is 328.98 mm, while for the HG PM with embedded PM rotor, it is 283.45 mm.

Table 5.23 summarizes the two designs. Note that the HG has a single stator shared between the PM and WF rotor sections. Therefore, the machine power is split between PM and WF by their respective voltage ratio since the stator current is the same for both sections. For both HG Design 1 and Design 2, the permanent magnet material used in the machine is NdFeB. Both designs have the same WF rotor; however, manufacturing the PM rotor is different in embedded and surface designs. The HG with embedded magnet PM rotor results in 14% lower axial length which in turn results in lower machine mass. The end winding for the WF rotor is extended by 20 mm beyond the WF axial length. Therefore, the stator axial length needs to be increased to account for the WF end winding extension as seen from Fig. 5.51. The HG axial length including the end winding extension is listed in Table 5.23.

The WF rotor of the finial HG is the same as the 9-phase HG WF with a scaled axial length. Therefore, the WF rotor winding resistance is calculated from Eqs. (5.10) to (5.14) and listed in Table 5.24. The HG stator is basically the same as stator for the 9-phase HG WF and HG PM with a scaled axial length. The stator

Table 5.23 Specification details for the HG designs

Parameter	WF section	PM section	HG
HG Design 1 with surface NdFeB PM rotor			
Output power (MW)	0.9	2.7	3.6
RMS phase voltage (kV)	2.7	8.1	10.80
Active axial length (mm)	63.25	265.73	328.98
Axial length with end winding extension (mm)	83.25	256.73	349
Number of turns per stator coil (p.u.)	19		
Coil cross section (mm^2)	380.64		
Turn cross section (mm^2)	20.03		
HG Design 2 with embedded NdFeB PM rotor			
Output power (MW)	0.9	2.7	3.6
RMS phase voltage (kV)	2.7	8.1	10.80
Active axial length (mm)	63.25	220.2	283.45
Axial length with end winding extension (mm)	83.25	220.2	304
Number of turns per stator coil (p.u.)	19		
Coil cross section (mm^2)	380.64		
Turn cross section (mm^2)	20.03		

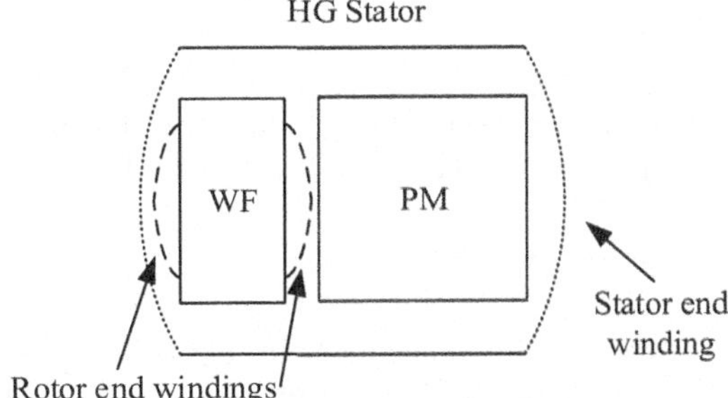

Fig. 5.51 HG rotor and stator end winding extension

Table 5.24 HG wound field rotor winding specifications

Item	Design 1 and Design 2
Rotor turn length (mm)	826.5
Rotor turn cross-section area (mm^2)	112
Rotor turn resistance (Ω)	0.000124
Number of rotor coil turns	51
Rotor coil resistance (Ω)	0.0063
Rotor winding interconnection length (mm)	675.44
Rotor winding interconnection resistance (Ω)	0.0001
Rotor resistance (Ω) at 20 °C	0.064
Rotor resistance (Ω) at 75 °C	0.078

Table 5.25 HG stator winding specifications

Item	Design 1	Design 2
Stator turn length (mm)	2396.07	2306.07
Stator coil cross section (mm^2)	380.64	
Number of stator coil turns	19	
Stator turn cross-section area (mm^2)	20.03	
Turn resistance (Ω)	0.002	0.0019
Coil resistance (Ω)	0.038	0.036
Interconnection length (mm)	919.79	
Interconnection resistance (mΩ)	0.8	
Phase resistance (Ω) at 20 °C	0.58	0.55
Phase resistance (Ω) at 75 °C	0.70	0.67

Table 5.26 HG per-phase stator equivalent synchronous inductance

Item	Design 1	Design 2
Inductance due to WF rotor (mH)	44.53	
Inductance due to PM rotor with surface NdFeB (mH)	182.38	–
Inductance due to PM rotor with embedded NdFeB (mH)	–	581.26
Total stator per-phase (synchronous) inductance (mH)	226.91	625.79

per-phase winding resistance is presented in Table 5.25. The stator per-phase inductance is the sum of the inductances due to WF and PM sections. The inductance is a function of machine axial length and square of number of turns:

$$L \propto N^2 \, l_a \tag{5.28}$$

Therefore, the inductances for the HG WF and HG PM machines are scaled and added to obtain the HG per-phase inductance. Table 5.26 lists the HG equivalent per-phase synchronous inductance.

5.8.2 HG Loss Audit

The losses in the HG is split into stator and rotor copper loss, stator iron (or core) loss, and mechanical losses such as windage and friction. The copper losses for the stator and rotor windings are calculated via:

$$\begin{aligned} P_{\text{cu}}^s &= 9\, R_s\, I_s^2 \\ P_{\text{cu}}^r &= R_r\, I_r^2 \end{aligned} \tag{5.29}$$

where

R_s: Stator per-phase resistance at 75 °C (Ω)

Table 5.27 HG copper, friction and windage losses at nominal speed, full-load RMS stator current, and rotor full-field DC current

Item	Design 1	Design 2
Stator copper loss (kW)	91.7	87.80
Rotor copper loss (kW)	18.44	
Windage and friction loss (kW)	54.62	47.58

R_r: Rotor winding resistance at 75 °C (Ω)

I_s: Stator RMS full-load phase current (A)

I_r: Rotor DC field current at full-load (A)

Since the HG rotational speed and stator structure are the same as the benchmark SG, the windage and friction losses for the HG are scaled by the axial length from their original values for the benchmark SG. Table 5.27 details the HG copper, friction and windage losses at nominal speed (600 RPM), full-load RMS stator phase current (120.67 A), and rotor full-field DC current (486.23 A).

The estimation of iron losses in electric machines has been studied in many research publications [71, 74]. However, much of the work is linked back to the early work of Steinmetz [75] on the measurement and characterization of hysteresis loss in lamination steels. Some studies divide the iron losses into three main components, namely, hysteresis, eddy current, and excess losses [74]. Separation of the losses by linear magnetization, rotational magnetization, and higher harmonics is also studied among the loss analysis publications [73], while the losses calculated in [72] are proportional to the square of the rate of change of flux-density. Finite element analysis has been used to improve the loss predictions. In this book, the iron losses are predicted via polynomial regression which is used to fit the curve of actual iron loss data of the United Laminated steel material (0.47 mm, 26 gauge) [76] at different frequencies. The manufacturer's iron loss curves are shown in Fig. 5.52 where the core losses are measured in Watts per kilogram, that is, loss density (W/kg), at different peak flux-densities and for two frequencies, 50 Hz and 60 Hz.

A third-order polynomial equation is fit into the manufacturer's iron loss curves:

$$P_{\text{iron}}^d = k_e\, B_m^3 + k_h\, B_m^2 + k_a\, B_m + k_d \tag{5.30}$$

Second-order polynomial functions are used to estimate the coefficients in the loss formula:

$$\begin{aligned}
k_e &= a_{e2}\, f^2 + a_{e1}\, f + a_{e0} \\
k_h &= a_{h2}\, f^2 + a_{h1}\, f + a_{h0} \\
k_a &= a_{a2}\, f^2 + a_{a1}\, f + a_{a0} \\
k_d &= a_{d2}\, f^2 + a_{d1}\, f + a_{d0}
\end{aligned} \tag{5.31}$$

where.

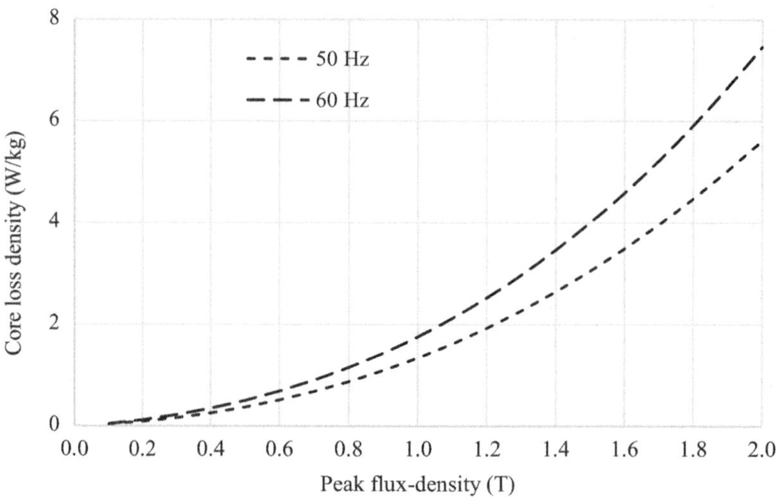

Fig. 5.52 Manufacturer's iron loss data for United Laminated steel

Table 5.28 Constants of the loss coefficients

Item	Value (p.u.)
a_{e2}	0.0001
a_{e1}	0.0041
a_{e0}	-0.2873
a_{h2}	0.00004
a_{h1}	0.0002
a_{h0}	0.8743
a_{a2}	-0.00001
a_{a1}	0.0259
a_{a0}	-1.0767
a_{d2}	0.000007
a_{d1}	-0.003
a_{d0}	0.148

P^{d}_{iron}: Core/iron loss density (W/kg)

B_m: Maximum/peak flux-density (T)

k_e, k_h, k_a, k_d: Frequency-dependent loss coefficients (p.u.)

a_{ei}, a_{hi}, a_{ai}, a_{di}: Constants of the loss coefficients where $i = 0$, 1, 2

The constants of the frequency-dependent loss coefficients are listed in Table 5.28 while Fig. 5.53 compares the predicted and manufacturer's iron loss density. As seen, the predicted loss curves fit the datasheet with good agreement. The machine frequency is calculated via:

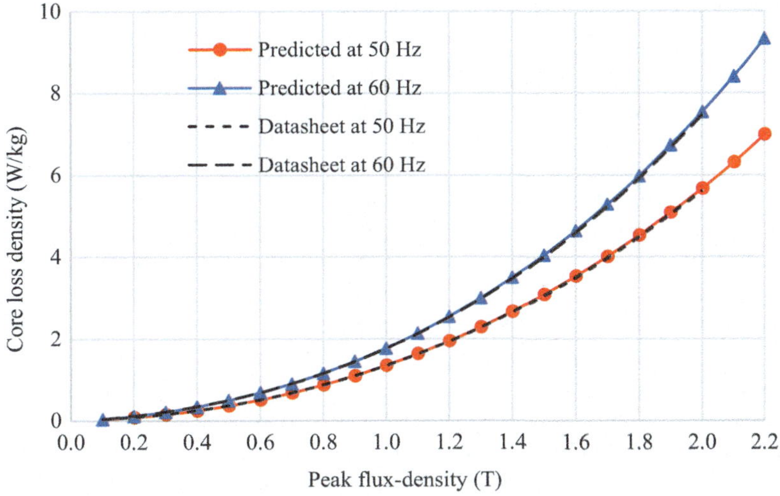

Fig. 5.53 Predicted versus manufacturer's datasheet iron loss density

Table 5.29 Flux-density for different sections for the HG

Section	Flux-density (T)		
	Surface NdFeB PM	Embedded NdFeB PM	WF
Stator tooth tip	1.04	1.3	1.77
Stator tooth body	1.27	1.54	2.0
Stator back-iron	1.44	1.81	1.8

$$f_s = \frac{P}{120}\, n_s \tag{5.32}$$

where

n_s: Rotational speed (RPM)
P: Number of poles

The HG nominal rated rotational speed is 600 RPM; hence, having 10 poles, the machine frequency is 50 Hz.

The stator tooth and back-iron operate at different flux-density as listed in Table 5.29. Therefore, iron losses for three sections, namely, stator tooth tip, stator tooth body, and stator back-iron, are calculated separately. The total iron loss is the sum of the losses for the three sections. The stator core losses for the PM and WF sections are calculated separately due to different flux-densities.

The iron losses calculated from Eq. (5.30) are per kilogram iron. The iron loss for each section is obtained via multiplying the section mass by the loss density. The mass of each section is calculated by using the United Steel Lamination density and multiplying it to the section volume. Table 5.30 shows the volume and mass of different section, while the HG iron losses for different sections are presented in

Table 5.30 Volume and mass audit of the HG designs

Item	Sections					
	Surface NdFeB PM		Embedded NdFeB PM		WF	
	Volume (m^3)	Mass (kg)	Volume (m^3)	Mass (kg)	Volume (m^3)	Mass (kg)
Stator tooth tip	0.0086	65.79	0.0072	55.08	0.0027	20.66
Stator tooth body	0.106	810.9	0.088	673.2	0.033	252.45
Stator back-iron	0.376	2876.4	0.313	2394.45	0.118	902.7
Rotor pole steel	–	–	–	–	0.0625	478.17
Rotor back-iron	0.2649	2026.3	0.628	4805.21	0.0713	545.71
Rotor pole PM	0.06859	524.69	0.0578	442.34	–	–
Stator copper	0.09452	846.92	0.08985	805.06	0.038	340.9
Rotor copper	–	–	–	–	0.0216	193.54
Total stator steel	0.49	3753.09	0.41	3122.73	0.15	1175.81
Total rotor steel	0.2649	2026.3	0.628	4805.21	0.1338	1032.88
Total copper	0.09452	846.92	0.08985	805.06	0.0595	533.54
Exciter	–	–	–	–	–	285.31
Total mass (tonnes)	6.63	8.73	3.03			
Section	HG					
	Design 1			Design 2		
	Volume (m^3)		Mass (kg)		Volume (m^3)	Mass (kg)
Stator tooth tip	0.0113		86.45		0.0099	75.74
Stator tooth body	0.193		1063.35		0.121	925.65
Stator back-iron	0.494		3779.1		0.431	3297.15
Rotor pole steel	0.0625		478.17		0.0625	478.17
Rotor back-iron	0.3362		2572.01		0.06993	5350.92
Rotor pole PM	0.06859		524.69		0.0578	442.34
Stator copper	0.13252		1187.82		0.12785	1145.96
Rotor copper	0.0216		193.54		0.0216	193.54
Total stator steel	0.64		4928.9		0.56	4298.54
Total rotor steel	0.3989		3050.18		0.13243	5829.09
Total copper	0.15412		1381.36		0.14945	1339.5
Brushless exciter	–		285.31		–	285.31
Total mass[a] (tonnes)	10.17				12.19	

Steel (lamination) density $=$ 7650 kg/m^3; copper density $=$ 8960 kg/m^3; PM mass density $=$ 7500 kg/m^3

[a]Excluding shaft, machine casing, end caps, cooling system, and other support structure

Table 5.31. Compared to the fully wound filed machine (i.e., SG), the HG Design 1 results in 3.9% lower mass, while the HG Design 2 results in 15.2% higher mass. This is due to the fact that rotor back-iron in the embedded PM design is increased. Note that mass of the shaft, machine casing, end caps, cooling system, and other

Table 5.31 HG iron losses

Item	Core loss (kW)				
	Surface NdFeB PM	Embedded NdFeB PM	WF	HG Design 1	HG Design 2
Stator tooth tip	0.10	0.13	0.09	0.19	0.22
Stator tooth body	1.78	2.19	1.43	3.21	3.62
Stator back-iron	8.16	10.98	4.09	12.25	15.07
Total	10.03	13.30	5.61	15.64	18.91

Table 5.32 HG total losses

Item	Design 1		Design 2		9-phase SG	
	(kW)	(%)	(kW)	(%)	(kW)	(%)
Stator copper	91.7	50.8	87.80	50.8	85.18	44.8
Rotor copper	18.44	10.2	18.44	10.7	26.7	14.8
Windage and friction	54.62	30.3	47.58	27.5	39.6	20.8
Iron	15.64	8.7	18.91	11	38.58	19.6
Total	180.4	100	172.73	100	190.1	100
Efficiency (%)	95.23		95.42		94.98	

support structures are not included in the analysis. The total losses for the HG for both Design 1 and Design 2 are listed in Table 5.32. For Design 1, the stator and rotor copper losses contribute to 50.8% and 10.2% of the total loss, the windage and friction contribute to 30.3%, and the iron loss is 8.7% of the total machine losses. For Design 2, the stator and rotor copper loss are 50.8% and 10.7% of the total loss while windage and friction is 27.5% and iron loss contributes to 11% of the total losses. The HG efficiency at 3.6 MW output for Design 1 and Design 2 is 95.23% and 95.42%, respectively, while the efficiency of fully wound field SG is 94.98% as seen from Table 5.32.

5.9 Summary

This chapter discusses design of the HG. The HG design is based on an industrial 3-phase SG referred to as benchmark SG. The benchmark machine is rewound for a 9-phase stator, and FEA analysis is carried out. It is shown that compared to the 3-phase, the 9-phase machine for the same ampere-turn, coil cross-section area, and stator copper loss results in 4.2% higher voltage and hence power. Performance of the machine connected to an active power electronic VSC with two control methods, PWM and overmodulated, is evaluated and compared to the case where the machine is connected to a passive rectifier. It is concluded that when a passive rectifier replaces a VSC with PMW or overmodulated control, the machine needs to be redesigned for 10% and 42% higher-rated power. Three designs for the PM rotor section is investigated: a surface PM rotor with NdFeB magnets, an embedded PM rotor with ferrite magnet, and an embedded PM rotor with NdFeB magnets. It is shown that due to machine geometry, the PM rotor with ferrite magnets is not a suitable option for the machine. The air-gap flux-density for the surface PM rotor is 0.842 T compared to 0.998 T for the WF rotor. However, using the embedded PM rotor, the air-gap flux-density is boosted to 1 T. The HG losses are split into stator and rotor copper, windage and friction, and iron losses. The benchmark SG windage and friction losses are scaled by the axial length ratio to obtain the values for the HG. The machine copper losses are calculated at rated stator current and full-field rotor current. The HG iron losses are estimated using polynomial equation fitted to the manufacturer loss datasheet. The iron losses are calculated for stator tooth tip, stator tooth body, and back iron separately and added up to obtain the total iron losses. For Design 1, the stator and rotor copper losses contribute to 50.8% and 10.2% of the total loss, the windage and friction contribute to 30.3%, and the iron loss is 8.7% of the total machine losses. For Design 2, the stator and rotor copper loss are 50.8% and 10.7% of the total loss, while windage and friction is 27.5%, and iron loss contributes to 11% of the total losses. The HG efficiency with 3600 kW output for Design 1 and Design 2 is 95.23% and 95.42%, respectively.

Chapter 6
High Voltage Insulation Systems

6.1   Introduction

The wind generation scheme in Chap. 3 uses a 9-phase HG with 38.1 kV RMS phase voltage. However, design of the HG in Chap. 5 was based on an SG with 6.35 kV RMS phase voltage (11 kV RMS line-to-line). Therefore, the HG windings specified in Chap. 3 need to be redesigned to accommodate a nominal phase RMS voltage of 38.1 kV. The first part of this chapter considers system analysis of the wind generation scheme using an HG with a phase RMS voltage of 6.35 kV, as previously designed from the benchmark SG, and then compares this machine with the results obtained at 38.1 kV in Chap. 3. The benchmark SG, an example of high-voltage machines in industry, has a phase voltage of 6.35 kV; hence, the system analysis at the machine voltage of 6.35 kV is carried out to verify the system performance, benefits, and gains at a voltage level already practiced by industry. The rest of this chapter discusses insulation system design for the HG windings suitable for a 38.1 kV RMS phase voltage machine. The machine initial 6.35 kV voltage is scaled in steps to obtain the targeted 38.1 kV and appropriate winding insulations for each design step is considered. Various industry practices for the HV windings are introduced, and insulation systems for different parts of a winding, that is, turn, strand, and ground insulations, are addressed in this chapter. Semiconductor and stress grading insulations are also investigated. The voltage stress between different insulations and the grounded machine core is modeled using lumped capacitors to illustrate the voltage withstand strength of the insulations.

© Springer Nature Switzerland AG 2020
O. Beik, A. S. Al-Adsani, *DC Wind Generation Systems*,
https://doi.org/10.1007/978-3-030-39346-5_6

6.2 Analysis of the HV System Employing a 6.35 kV HG

For an RMS phase voltage of 6.35 kV (peak voltage of 8.98 kV), the rectified DC voltage at the output of the 9-phase HG calculated from Eq. (3.17) is 17.6 kVDC. This voltage is one-sixth of the DC voltage for the original system, that is, 105.6 kVDC. The voltage ratio is indeed the ratio of the RMS phase voltages (6.35/38.1 = 1/6). For a 3.6 MW nominal turbine power, the DC current at the rectifier output is 204.55 A, that is, six times the currents in the original system. The DC voltage for the transmission cable connecting the offshore to the onshore substations is the same as the original system, that is, 420 kV. Therefore, the DC/DC converter step-up ratio is increased to 23.86 (420/17.6). Assuming the same current densities for the cables as in the Walney system discussed in Chap. 3, the cable cross sections and resistances are, respectively, six and one-sixth times than their respective original values. Therefore, the resistances are reduced by a factor of one-sixth except for the transmission cable resistance. Table 6.1 lists the related cable cross sections, current densities, and resistances for the system operating with a 6.35 kV HG. The results of the system analysis at full load and with a 6.35 kV HG are presented in Table 6.2.

Table 6.3 presents a comparison between the two systems, that is, system with a 6.35 kV HG and with a 38.1 kV HG at full load. It is seen that the total system efficiency is similar for both systems resulting in similar benefits and gains in the new system. For the 38.1 kV HG, the mass is increased by 7.5% due to increase in the machine outer diameter; this will be discussed in detail at the end of this chapter after insulation considerations. However, the total mass for the system with 6.35 kV HG is 40.9% higher than the system with 38.1 kV HG.

These results form an interesting conclusion. One of the objectives of the HV HG wind generation scheme is to remove active power electronic converters from the

Table 6.1 System cable specifications with a 6.35 kV HG

Description	Tower	Circuit 2		Offshore substation to onshore
		Inter-array between turbines	Inter-array to offshore substation	
Cable length (km)	0.0835	0.750	0.458	44.4
Material	Copper			
Copper resistivity ($\mu\Omega$m)	0.0172			
Rated voltage (kVDC)	20	20	20	425
Current density (A/mm^2)	1.58	1.58	0.6	1.2
Cross-section area (mm^2)	129.48	517.86	1705.2	362.31
DC cables resistance (Ω)	0.022	0.05	0.0092	4.2

Table 6.2 System full-load analysis with a 6.35 kV HG

Turbine					
Turbine no.	1	2	3	4	5
HG phase voltage (kV)	6.35	6.35	6.34	6.33	6.33
HG DC voltage (kV)	17.60	17.59	17.57	17.54	17.54
HG DC current output (A)	197.41	197.57	197.79	198.13	198.13
DC voltage at the tower base, kV	17.60	17.59	17.57	17.54	17.50
Interconnection and transmission					
Input DC voltage at the offshore substation (kV)	17.49				
Input DC current, Circuit 2 at offshore sub. (A)	989.04				
Output DC voltage at the offshore sub. (kV)	417.27				
Output DC current at the offshore sub. (A)	416.57				
Input DC voltage at the onshore sub. (kV)	415.52				
Input DC current at the onshore sub. (A)	416.57				
Power and losses in the nacelle and tower					
Turbine power (kW)	3600				
HG output power (kW)	3492				
Rectifier output power (kW)	3474.54				
HG efficiency (%) and losses (kW)	97	108			
Rectifier efficiency (%) and losses (kW)	99.5	17.46			
Tower cable power loss (kW)	0.86				
Turbine link cable					
From–to		1 to 2	2 to 3	3 to 4	4 to 5
Inter-array cable loss (kW)		1.95	7.80	17.57	31.28
Cable loss, Circuit 2 to the offshore substation (kW)		9.00			
Input power and losses at offshore sub. from Circuit 2					
Power (kW)	17,300.80	Losses (kW)	699.20		
Input power and losses at offshore sub. from Circuits 1–6					
Power (kW)	176,468.17	Losses (kW)	7131.83		
DC/DC converters efficiency at offshore sub. (%)	98.5				
Loss at offshore substation (kW)	2647.02				
Cable losses, offshore to onshore sub. (kW)	728.83				
Total power at onshore substation (kW)	173,092.31				
Total losses at onshore substation (kW)	10,507.69				
Total system efficiency (%)	94.28				

turbine towers. Effectively, the individual tower power electronic converters are rationalized into larger DC/DC converters that can be located at the central offshore substation. These HV DC/DC converters are still the subject of research and development. However, it is assumed here that they will be realized in due course as a natural progression of HVDC subsystems. Certainly, for HVDC transmission, high step-up, step-down ratio DC/DC converters are a technical requirement for future systems. Consequently, the results presented in Tables 6.2 and 6.3 illustrate the HV HG system concept could be technically realized at an accepted industry

Table 6.3 Comparison between systems with a 6.35 kV HG and with a 38.1 kV HG at full load

Items	System	
	6.35 kV HG	38.1 kV HG
Number of HG phases	9	
HG RMS phase voltage (kV)	6.35	38.1
Rectified DC voltage (kVDC)	17.6	105.6
Turbine power (kW)	3600	
System total losses (kW)	10,507.69	9894.10
Total system efficiency (%)	94.28	94.61
DC/DC step-up ratio (p.u.)	23.86	4
Active power conversion mass (tons)		
HG	24.2	26.02
Tower cable	0.18	0.03
Inter-array cables between turbines	72	12
Cables from circuits to offshore substation	59.34	9.89
Transmission cable	215.33	
Total cable	346.85	237.25
Total system mass	371.05	263.27

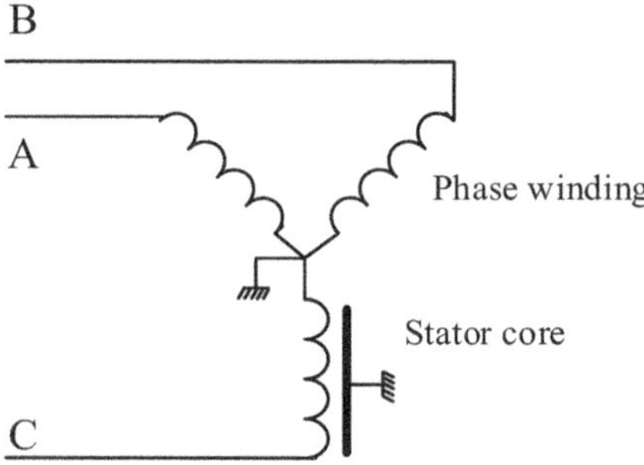

Fig. 6.1 Stator winding connections

standard voltage level (11 kV line-to-line), albeit with higher interconnecting cable mass and higher step-up ratio for offshore DC/DC converter. This being accepted, the challenge herein is to therefore investigate the potential design changes necessary to increase operating voltages from 11 kV line-to-line, or 6.35 kV phase up to 38.1 kV phase. Note, phase voltages are quoted here since there are four different line-to-line voltages specified for a 9-phase system.

6.3   HV Winding Types

Figure 6.1 schematically shows a 3-phase HG stator winding connections. Each phase is connected to the neutral point in one end and to the machine terminals at the other. The neutral point is solidly connected to the surrounding ground along with

Fig. 6.2 A random-wound stator

the stator core. This provides a level of equipment protection as well as referencing the winding coils to ground and the stator core, which is also grounded.

The winding insulations become more significant as machine nominal voltage increases. Insulation does not contribute to the machine torque production [77]; however, it is required to prevent short circuits between the copper conductors and also between conductors and ground. Insulations also provide mechanical strength and help protect conductors from damage in the process of coil insertion into the stator slots and also prevent the coils from movements during operation. Main types of stator windings are:

- Random-wound coils
- Form-wound coils
- Roebel bar coils

In random-wound coils, the insulated conductors are usually round cross-section wires that are randomly and continuously wound to form a multi-turn coil. In this type of winding, it does not matter if the turn with the highest voltage is placed adjacent to a turn with the lowest voltage. This type of winding is not used for high-voltage applications. Random-wound stators could have coils wound beforehand and inserted into slots or wound directly into the coils around stator teeth. Figure 6.2 shows a random-wound stator winding for a low-voltage (300 V) machine.

Form-wound coils consist of individually insulated copper wires or rectangular bars that are wound and preformed to a fixed shape for subsequent insertion into the machine stator or wound rotor. The rectangular shape allows better packing factor for the slots. In form-wound coils, each turn consists of a few to several copper strands. Figure 6.3 shows a form-wound stator coil.

Form-wound coils are used for higher-voltage machines. Unlike random-wound coils, care has to be taken when designing form-wound coils such that each turn is next to a turn with least voltage difference. This results in minimized insulation between turns.

Roebel bar is another windings scheme, as shown in Fig. 6.4, that is sometimes used for high-voltage and high-power machines such as turbo generators over

Fig. 6.3 A form-wound coil

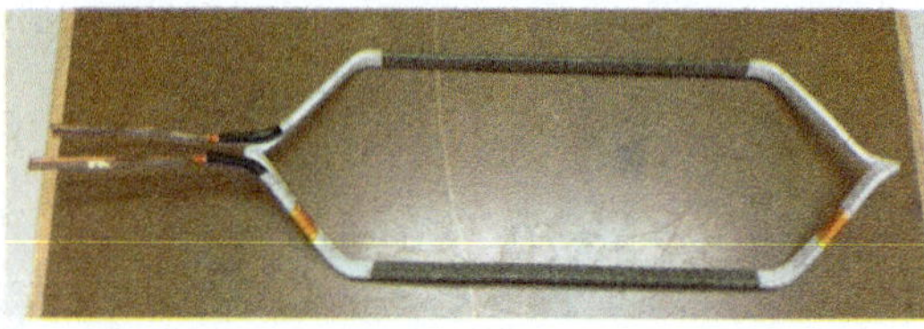

Fig. 6.4 A typical Roebel bar winding (Photo courtesy Partzsch Group) [79]

50 MW [78]. The Roebel bar winding implementation transposes the winding turns to minimize parasitic winding effects such as eddy current and proximity effects, cross-slot leakage, unbalanced turn voltage, etc. In this book, form-wound coils are considered for the HV HG windings; hence, other types of windings will not be studied.

6.4 Insulation Systems

In order to have a sensible design for the HG insulation systems that is comparable and verifiable with industry practice HV windings, the 11 kV insulations for the benchmark SG (3-phase HG WF) are used as a design base, against which new insulations systems are compared. Figure 6.5 shows the cross section of a stator slot for the 3-phase benchmark SG. The slot has two layers, each representing the cross section of one coil side. The two coils are from either the same phase or different phases. Therefore, the voltage difference between the two layers is different depending on the coil voltages. Each coil has four turns, each turn of which is referred to as conductor; each conductor has six strands that are transposed from coil to coil and connected at the machine terminals in parallel. There is no or may be small voltage difference between the strands; therefore, they require minimum insulation.

Strands have an insulation thickness of 0.05 mm, while conductor insulation thickness is 0.1 mm. Top and bottom coils are separated from each other in the slot by a 4 mm thick insulation. The slot has a wedge made of insulating material of 5 mm thickness. The slot wedge helps to keep the coils tightly placed in the slot and prevents their movements. The laminations are grounded, and between the slot wall

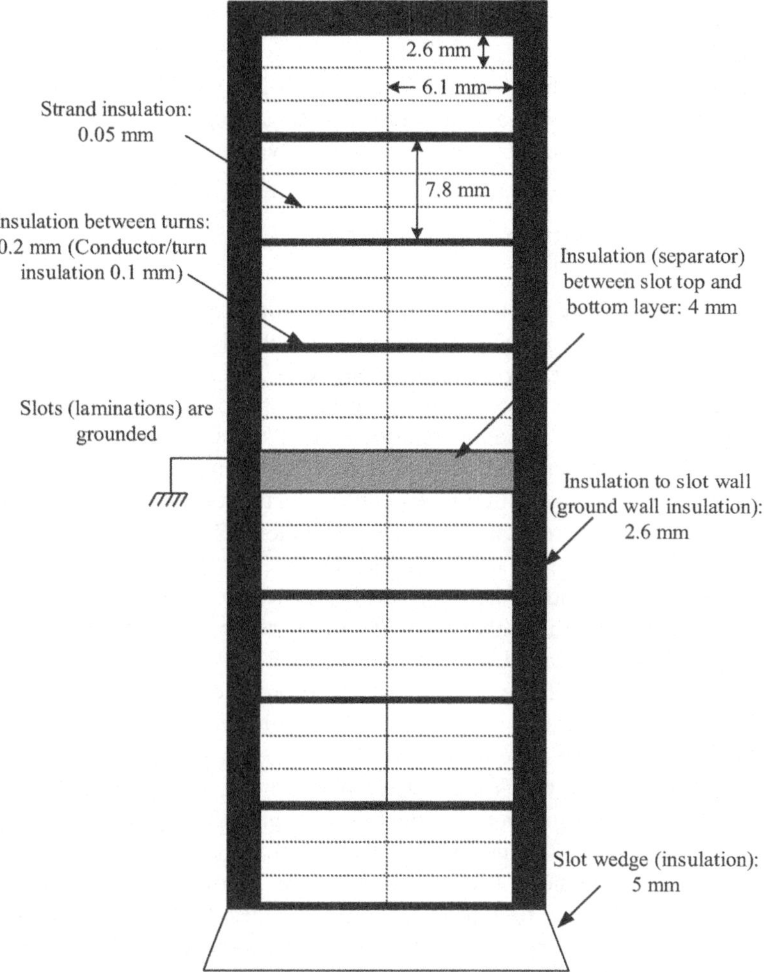

Fig. 6.5 Three-phase benchmark SG or HG WF slot insulation system

and coils, there is an insulation of 2.6 mm thickness; this is referred to as groundwall insulation. Main insulation components in a HV winding include:

- Strand insulation
- Turn insulation
- Groundwall insulation
- Semiconductive slot insulation
- Voltage stress grading insulation

6.4.1　Strand, Turn, and Groundwall Insulation

Each conductor in the 3-phase HG WF has 6 strands that are formed and electrically connected in parallel to form a single turn. Splitting a big conductor cross section into smaller strand cross sections has three main advantages:

- Mechanical and manufacturing ease
- Reduction of eddy current effects and losses
- Reduction of copper losses

Mechanically, a conductor with large cross section is difficult to bend and insert into slots. Forming a conductor from smaller strands facilitates bending and manufacturing. Moreover, by reducing conductor cross-section area to smaller strands, the skin effect is reduced providing the conductor strands are transposed when interconnecting the individual phase coil. Unlike DC current that is uniformly distributed in the conductor cross-section area, AC current tends to flow in the periphery of a conductor resulting in a smaller effective cross section which results in a perceived higher AC resistance than the conductor DC resistance. The higher the conductor resistance, the higher the copper losses. By splitting a conductor into smaller cross sections and transposing at inter-coil connections, the skin effect is reduced, and hence more of the available conductor cross section is used. The 3-phase HG WF is rated at 11 kV line-to-line RMS voltage (6.35 kV RMS phase voltage). For each phase, there are 45 coils; therefore, the individual coil voltage is 140 V. Each coil has four turns; therefore, the turn voltage is 35 V. Each turn is made up of six parallel strands; hence, the voltage across each strand is the same as turn voltage, as shown in Fig. 6.6. Table 6.4 summarizes coil, turn, and strand voltages.

Turn insulations prevent short circuits between turns that are in contact. In order to minimize the turn insulations, each turn in a coil is placed next to the turn with smallest voltage difference. Turn insulations are normally thicker than strand insulations since they have to withstand higher voltages. However, some manufacturers combine turn and strand insulations to facilitate manufacturing process. The turn voltage buildup for the 3-phase HG WF is illustrated in Fig. 6.7a.

Fig. 6.6 Three-phase HG WF strand voltage

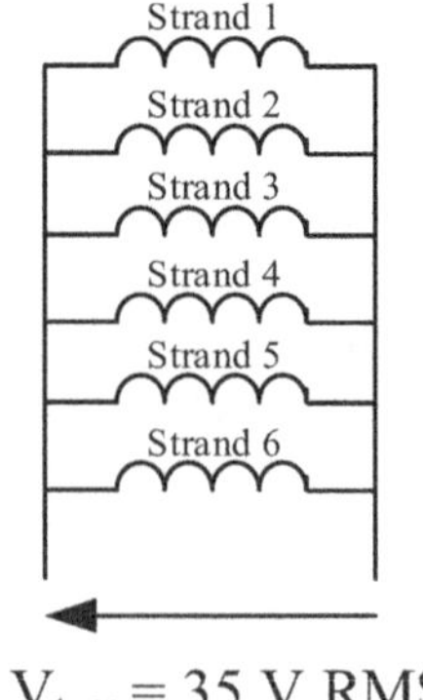

Table 6.4 Turn, coil, and strand voltages

Description	RMS	Peak
Line-to-line voltage (kV)	11	15.56
Phase-to-ground voltage (kV)	6.35	8.98
Voltage across each coil (V)	140	200
Voltage across each turn (V)	35	50
Voltage across each strand in each turn (V)	35	50

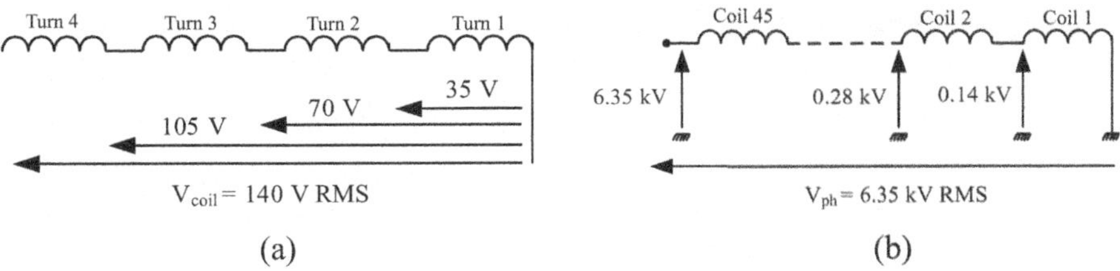

Fig. 6.7 Three-phase benchmark SG (or HG WF) turn and phase voltages (**a**) Turn voltage buildup (**b**) Phase voltage buildup

Once the turns are placed together and a coil is formed, a layer of insulation referred to as groundwall insulation, usually a tape, is applied on the coil surface. The groundwall insulation protects coils from contact to the grounded laminations. It also holds the turns tightly together and prevents them from moving due to the forces in the machine acting on conductors. For random-wound windings, the turn insulation also acts as groundwall insulation. Therefore, the turn insulation in random-wound coils is designed to withstand phase-to-phase voltages. However, for form-wound coils in high-voltage machines, groundwall insulation is normally different from turn insulation. The coils, depending on their position in a phase, experience different voltages with respect to the grounded laminations. The first coil in a phase, that is, the coil that is connected to the ground, referred to as neutral-end coil, is subject to zero voltage, and as the coils are connected in series, their voltage to ground increases. Therefore, the last coil in a phase, referred to as the terminal-end coil is subject to full phase-to-ground voltage. Therefore, the groundwall insulations need to protect the coils from voltages as high as phase voltage against the grounded laminations. Although different coils experience different voltages, the groundwall insulations are normally designed for the same insulation thickness for all coils. Figure 6.7b shows the phase voltage buildup for the benchmark SG (or 3-phase HG WF).

A design for the HV windings for the ABB Powerformer was proposed where coils have different insulation thickness depending on their voltage, and hence, the slot size varies as the insulation thickness increases [80]. The coil insulation has to be applied coil-by-coil as the machine is wound, making the winding process somewhat involved. Figure 6.8 shows the machine stator slots, highlighting the variation in slot shape to accommodate the "cable" winding [80].

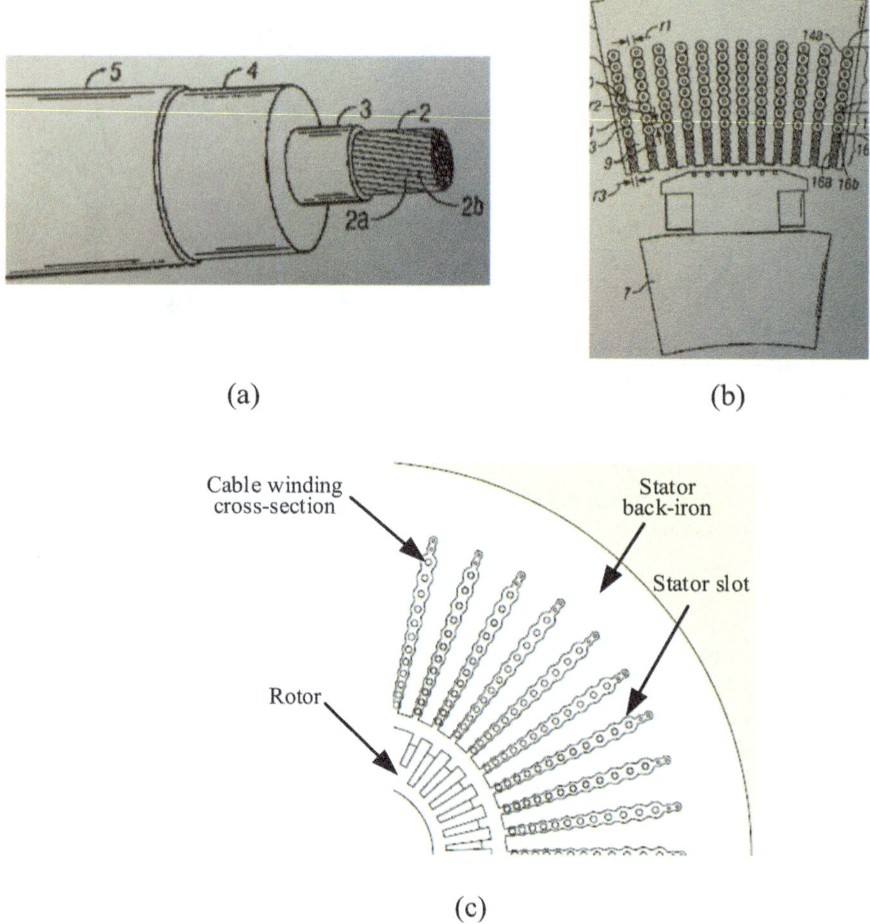

Fig. 6.8 AB Powerformer stator slots [80]. (**a**) Cable structure (**b**) Slot/tooth cross section (**c**) Machine cross section

Air pockets (also known as voids) can occur in the insulation system during the manufacturing process and when the machine is in service. If air pockets are subject to high voltages, they break down and a spark or discharge is initiated. Repeated electric breakdowns, referred to as partial discharge (PD), leave a hole in the insulation system; hence, the copper becomes exposed to the grounded laminations or other conductors and a short circuit ultimately occurs. The electric breakdown depends on the strength of the electric field, also referred to as electric stress or voltage gradient. The electric field between two surfaces displaced at a distance of d and with a voltage difference of V is:

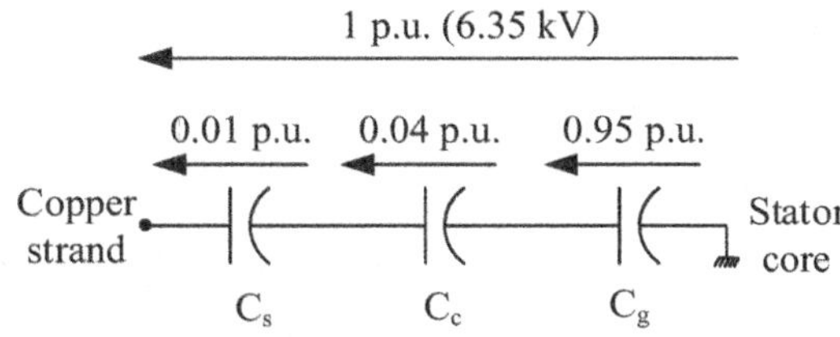

Fig. 6.9 Lumped capacitance circuit model between strands and grounded stator core

Table 6.5 Strand-to-groundwall capacitors parametric values

Symbol	Description	Parametric value
C_s	Strand insulation capacitor	$C_s = \frac{\varepsilon A}{0.05} = 20\,\varepsilon A$
C_c	Conductor insulation capacitor	$C_c = \frac{\varepsilon A}{0.1} = 10\,\varepsilon A$
C_g	Groundwall insulation capacitor	$C_g = \frac{\varepsilon A}{2.6} = 0.38\,\varepsilon A$

$$E = \frac{V}{d}\,(\text{kV/mm}) \tag{6.1}$$

If the voltage across two surfaces is increased, above breakdown point conduction occurs. In such case, the material between the surfaces (which could be air or an insulation material) is ionized and conducts current which ultimately creates a short circuit. For air, the breakdown strength at room temperature, low humidity, and one atmosphere is 3 kV/mm.

In order to evaluate voltage across insulations, each section of insulation is modeled as a capacitor circuit. The capacitance between two surfaces is:

$$C = \frac{\varepsilon A}{d}\,(\mu F) \tag{6.2}$$
$$\varepsilon = \varepsilon_r\,\varepsilon_0$$

where

ε Insulating material permittivity (F/m)
ε_r Insulating material relative permittivity
ε_0 Air permittivity (8.854×10^{-12} F/m)

A lumped capacitance circuit model for the capacitors between copper strands to the grounded stator core for the HG winding is presented in Fig. 6.9. There are three sections between the copper strand and grounded stator core:

(i)   Strand insulation modeled as C_s
(ii) Conductor insulation modeled as C_c
(iii) Groundwall insulation modeled as C_g

Using the insulation thicknesses for the 3-phase HG WF shown in Fig. 6.5 and Eq. (6.2), the capacitors parametric values are calculated and detailed in Table 6.5.

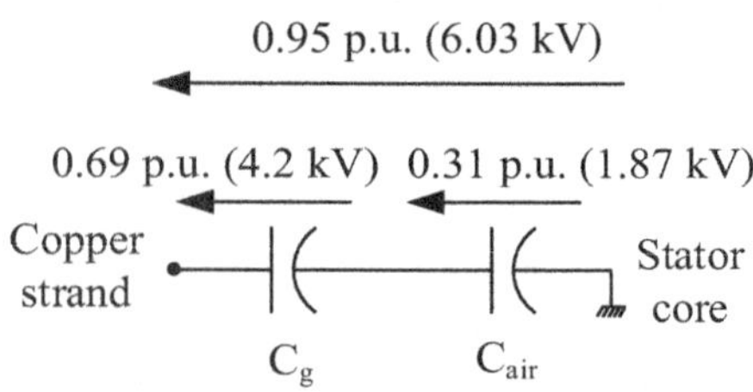

Fig. 6.10 Groundwall-to-grounded stator core model with an air pocket

As discussed, strands of the terminal-end coil are subject to the highest voltage, that is, 6.35 kV phase RMS voltage noted as 1 p.u. voltage. This voltage is divided between capacitors based on their capacitance as shown in Fig. 6.9. It is seen that the voltage across the groundwall insulation is highest due to its lowest capacitance. In order to study insulation failure, it is assumed there is an air pocket with 0.3 mm thickness in the groundwall insulation. The air pocket is modeled as a capacitor, C_{air}; hence, the circuit model from the groundwall to the grounded insulation is shown in Fig. 6.10.

Groundwall insulating material used for the machine has a permittivity of four times larger than air; therefore:

$$C_{air} = \frac{A}{d_{air}} = \frac{A}{0.3} = 3.33\,A$$

$$C_g = \frac{4A}{d_g} = \frac{4A}{2.6} = 1.54\,A$$

The total voltage between the groundwall insulation and grounded stator core is 6.03 kV (0.95 p.u.) as calculated in Fig. 6.9. This voltage is split between the groundwall insulation and air pocket based on their capacitance. Therefore, the voltage across the air pocket is 1.87 kV as shown in Fig. 6.10. The electric field across the air pocket is:

$$E_{air} = \frac{1.87}{0.3} = 6.23\,\text{kV/mm}$$

Air breaks down at 3 kV/mm; therefore, the air pocket breaks down and a discharge occurs in the groundwall insulation. In order to prevent this, air pockets are avoided during the manufacturing process by careful procedures in vacuum pressure impregnation (VPI) of the machine stator with insulating resins post winding. The HG stator has two coils in each slot. Figure 6.11 shows a capacitor circuit model for the insulation between the two coils. Insulations between the two coils include strand, conductor, and groundwall insulation. The separator between the two coils in the slot shown in Fig. 6.5 is modeled as C_{sep}, where $C_{sep} = \frac{\varepsilon A}{4} = 0.25\,\varepsilon A$.

The voltage difference between the two coils is as high as the line-to-line voltage if terminal-end coils of two different phases are placed in one slot. The voltages

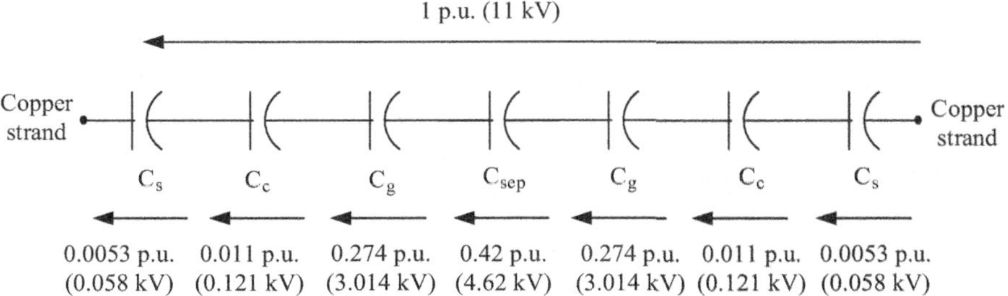

Fig. 6.11 Coil-to-coil capacitor core model

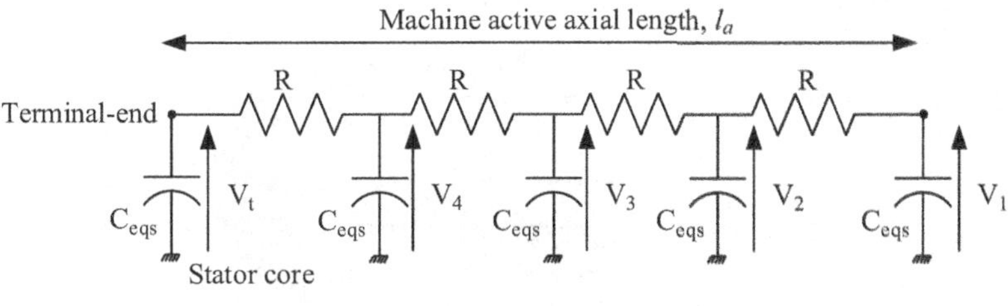

Fig. 6.12 Coil capacitor model along the HG axial length

across insulations are split based on their capacitance. As shown in Fig. 6.11, the highest voltage appears across the slot separator. Fig. 6.12 illustrates a circuit model for a coil along the axial length of the HG. Each capacitor in Fig. 6.12 is an equivalent capacitor ($C_{eqs} = 0.36\ \varepsilon A$) for three series capacitors C_s, C_c, and C_g. The winding axial length has been divided into four arbitrary sections each with the same resistance. The voltage profile at each point along machine winding axial length is shown in Fig. 6.12. As seen, the terminal end has the highest voltage.

6.4.2   Semiconductive Slot and Voltage Stress Grading Insulation

For the 3-phase HG WF, the copper packing factor is 0.457 p.u. Note that there are other definitions for copper packing factor; hence, copper packing factor is defined here as total slot copper area divided by total slot area. The total slot area is always larger than the total winding cross-sectional area including the coil insulations to facilitate practical placement of the coils. Therefore, there is an air gap between coil surface and the grounded slot. At high voltages, a PD can occur in this air gap for the same reason that a PD occurs in an air pocket in the insulation if not properly insulated. In order to fill the air gaps, some manufacturers use varnish/resin applied via VPI [81–84]. However, if the voltage across the air gap exceeds 3 kV/mm, the air

Fig. 6.13 Semiconductive resistance and air-gap capacitor model

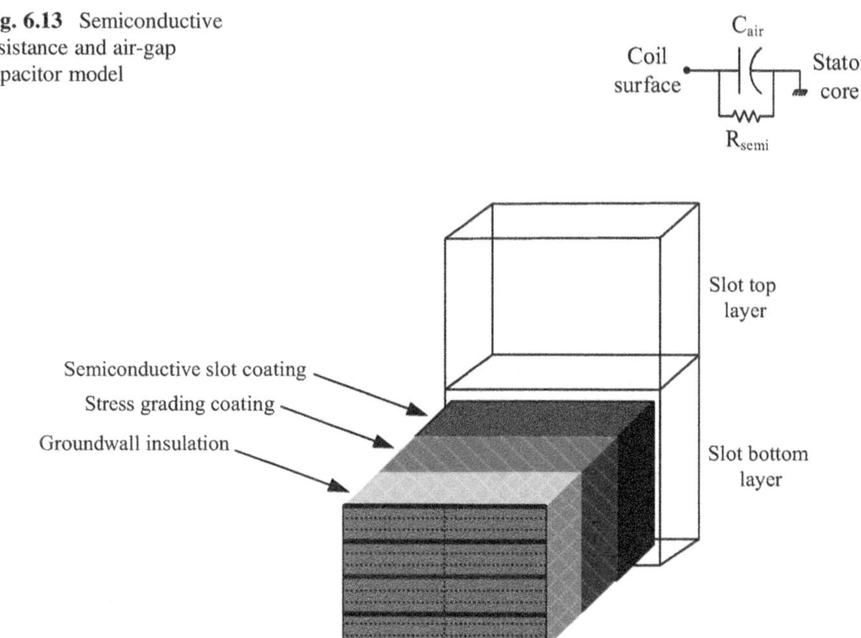

Fig. 6.14 Semiconductive, stress grading, and groundwall insulations for the 3-phase HG WF stator coil

breaks down and a PD occurs. In order to prevent a PD between coil surface and grounded slot, a semiconductive coating is applied on the coil groundwall insulation. This is also known as semiconductive slot coating or sometimes called conductive coating/tape/material [85, 86]. The semiconductive coating has a low resistance, and it is at ground voltage since it is in contact with the grounded slots. Semiconductive coating essentially puts a conductive layer in the air gap; hence, a low electric field is created and no PD occurs. A circuit model for the semiconductive resistance, R_{semi}, in parallel with the air-gap capacitor is shown in Fig. 6.13.

Semiconductive coating is not generally used if the machine is rated at voltages below 6 kV because the electric stress for such voltages will not exceed the 3 kV/mm limit. Higher conductivity coatings are not usually used because it short circuits the laminations and will result in eddy currents. Semiconductive coating is usually extended a few tens of millimeters from the end of the slot to neglect any end effects. Note that semiconductive coating should not be extended close to the end of the coil because the grounded surface will be too close to the connection of the coil to the next coil. Also, if the end winding is grounded by semiconductive tape, chances of a PD increases in the air pockets in the end winding as result of high electric field due to bringing the ground close to air pocket. Semiconductive insulations come usually as paint or tapes. Semiconductive slot insulations and groundwall insulation for the 3-phase HG WF stator coil are shown in Fig. 6.14.

The semiconductive coating cannot end abruptly since the voltage difference between the coils surface and grounded semiconductive tape creates an electric field that could exceed 3 kV/mm and starts a PD [85, 86]. Therefore, either the end of semiconductive coatings is terminated or, in most cases, a voltage stress grading material is continued to cover the end winding. The voltage stress grading coating is started at the end of the semiconductive slot coating, and often, it overlaps with the semiconducting slot coating. The stress grading coating is used to create a smooth transition from the grounded semiconductive slot coating to the high-voltage surface of the coil. The voltage stress grading material has lower conductivity than the semiconductive slot coatings. Silicon carbide is a common material that is used as stress grading material. The stress grading insulation comes usually as tapes. The stress grading insulation for the 3-phase HG WF stator coil is shown in Fig. 6.14.

6.4.3   Transposition

As discussed previously, turns in a coil are often transposed to reduce eddy current losses and also gain mechanical strength. The stator cross section for the coil start and coil end is shown in Fig. 6.15. Turn 1 is at the bottom of the start side of the coil but is transposed to the top at the end side. Turn 4 at the coil start side is the bottom turn, while Turn 4′ is transposed to the end side. Other turns in the coil also change places at the coil end side. This procedure is done for all the coils in each phase. In

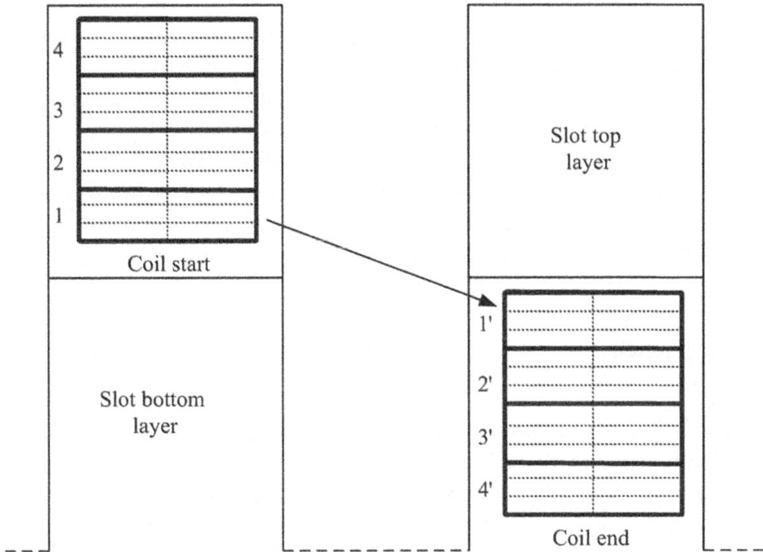

Fig. 6.15 Coil transposition

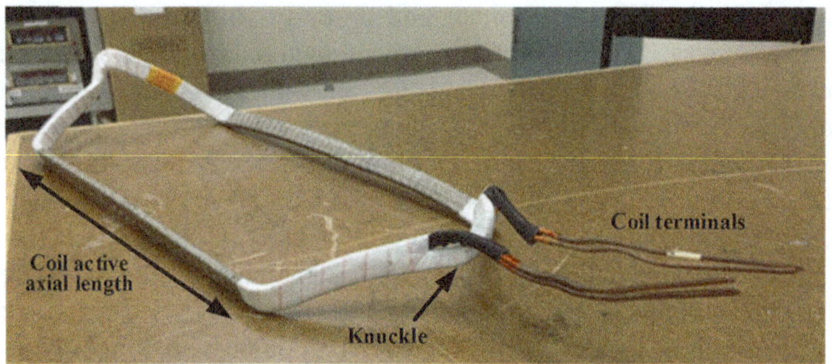

Fig. 6.16 Coil transposition via knuckle

Table 6.6 RMS phase, coil, turn, and strand voltage

Description	3-phase SG	9-phase HG
No. of coils	45	15
No. of turns	4	19
No. of strands per turn	6	2
No. of strands per coil	24	38
Phase voltage (kV)	6.35	6.6
Coil voltage (kV)	0.14	0.44
Conductor/turn voltage (kV)	0.035	0.023
Strand voltage (kV)	0.035	0.023

order to change the location of turns, the coil end winding is twisted, referred to as transposition via knuckle, as shown in Fig. 6.16.

In the transposition, turns with lowest voltage difference are placed next to each other. For instance, if the turn voltage is 0.25 p.u. (coil voltage 1 p.u.) and if Turn 4 is next to Turn 1, the voltage between them is 0.75 p.u., but if turns are arranged as shown in Fig. 6.15, the voltage between each consecutive turn is 0.25 p.u. and would hence require thinner insulation. Sometimes, supporting materials are used to mechanically support end windings for twisting actions, for instance, a cloth tape as shown in Fig. 6.16. The above discussed is a basic transposition via knuckle to reduce the eddy current loss and circulating current loss of the conductor.

6.5 HG Insulation

The 9-phase HG stator structure is the same as the benchmark 3-phase SG. However, 9-phase winding has 15 coils per phase as compared to 45 coils per phase for the 3-phase winding. The total slot copper cross section for the 9-phase HG winding is the same as for in the 3-phase SG. However, the number of turns per coil has increased to 19, reducing the turn voltage and cross-section area. Table 6.6 shows the RMS phase, coil, turn, and strand voltage for the 3-phase SG and the 9-phase

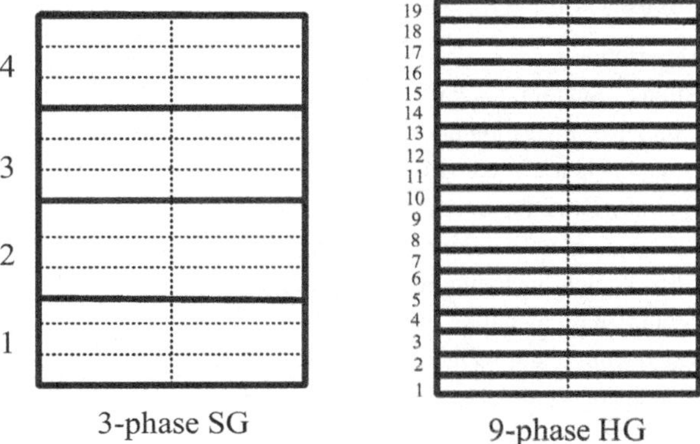

Fig. 6.17 Coil cross section for 3-phase SG and 9-phase HG

HG. Figure 6.17 illustrates the coil cross section for 3-phase SG and 9-phase HG. The strands are shown as dotted lines, while solid lines define turns. Each turn in the nine-phase system has two strands compared to six strands per turn for the 3-phase system. The strands and conductors in the 9-phase coil have the same width but reduced height. However, the voltage across strands and conductors in the 9-phase system is reduced by a factor of 0.66 (=0.023/0.035). Therefore, assuming that the strand and conductor insulation thickness vary linearly with voltage, the 9-phase strand and conductor insulations are 0.033 mm and 0.066 mm, respectively. However, it is assumed that the groundwall insulation thickness does not vary with the voltage; hence, at higher phase voltages, a better groundwall insulation material is used.

For the benchmark SG, the strand copper height and width excluding insulations are 2.6 mm and 6.1 mm, respectively, that is, strand cross-section area is 15.85 mm^2. There are 24 strands per coil; hence, the coil copper cross section is 380.64 mm where the coil copper width is 12.2 mm (2 strands) and coil copper height is 31.2 mm (12 strands). Having the same coil copper width and hence height for the HG and 19 conductors/turns per coil, the conductor copper height is 1.64 mm. The HG conductor copper and coil copper width are the same as the benchmark SG coils. However, since the insulation thickness for strands and conductors is changed, the total coil dimension is changed. Table 6.7 lists coil, conductor, and strand specifications for 9-phase and 3-phase coils. Compared to the benchmark SG, the HG total coil height is increased by 4.58%, but its width is decreased by 0.39%. The total coil cross section is, however, increased by 4.17%.

In order to minimize the voltage difference between consecutive turns (conductors), they are placed on top of each other. In the arrangement shown in Fig. 6.17, the voltage difference between two adjacent turns in the HG coil is 0.023 kV. However, if Turn 6 was to be placed next to Turn 1, the voltage difference would be 0.115 kV, and thicker conductor insulation would be required. Therefore, in designing the high-voltage coils, proper arrangement of turns needs to be taken into account.

The RMS phase voltage of the HG is 6.6 kV, whereas the wind generation scheme requires an HG with RMS phase voltage of 38.1 kV. Table 6.8 lists steps

Table 6.7 Coil, turn, and strands for 9-phase HG with phase RMS voltage 6.6 kV

Description	3-phase benchmark SG	9-phase HG
Strand insulation thickness (mm)	0.05	0.033
Conductor insulation thickness (mm)	0.1	0.066
Strand height excluding insulation (mm)	2.6	1.64
Strand width excluding insulation (mm)	6.1	6.1
Coil copper cross section (mm^2)	380.64	
Strand height including insulation (mm)	2.7	1.71
Strand width including insulation (mm)	6.2	6.2
Conductor height including insulation (mm)	8.3	1.84
Conductor width including insulation (mm)	12.6	12.53
Groundwall insulation thickness (mm)	2.6	
Coil height including insulation (mm)	38.4	40.16
Coil width including insulation (mm)	17.8	17.73
Total coil cross section including insulation (mm)	683.52	712.04

Table 6.8 Voltage scaling steps for the 9-phase HG

Description	RMS phase voltage (kV)	Step voltage multiplier
Initial phase voltage	6.6	–
Step 1	13.2	2
Step 2	26.4	2
Step 3	38	1.44

in which the machine voltage is increased from its initial 6.6 kV to obtain the desired voltage. In the first step, the initial 6.6 kV voltage is doubled, that is, 13.2 kV. Therefore, with the same number of coils, that is, 15, the coil voltage is doubled, that is, 0.88 kV from its initial 0.44 kV. Therefore, maintaining the same voltage per turn, that is, 0.023 kV, the number of turns need to be doubled to 38 turns per coil. By maintaining the same voltage per turn, the conductor insulation thickness is the same and hence a verifiable design.

Each turn in the 6.6 kV machine has two parallel strands which make up to a total of 38 strands per coil as shown in Fig. 6.17. By insulating strands using conductor insulation, each strand is used as one turn and the coil voltage is increased to 0.88 kV; hence, the machine voltage increases to 13.2 kV. The coil cross section for the 13.2 kV machine is shown in Fig. 6.18a.

In the 13.2 kV HG coil, the voltage difference between vertically adjacent turns is 0.046 kV (2×0.023 kV), while the maximum voltage difference between adjacent turns is 0.069 kV (3×0.023 kV). For the initial 6.6 kV machine, the conductor insulation is 0.066 mm. Therefore, for the 13.2 kV machine, the conductor insulation is increased to 0.13 mm, and a layer of insulation with a thickness of 0.13 mm needs to be added in the middle of the coil to account for maximum 0.069 kV voltage difference between turns. Table 6.9 shows the size of coil and turns for the 13.2 kV machine. As seen, the coil height and width are increased by 2.83% and 1.8%, respectively. The coil total cross-section area is, however, increased by 4.7%.

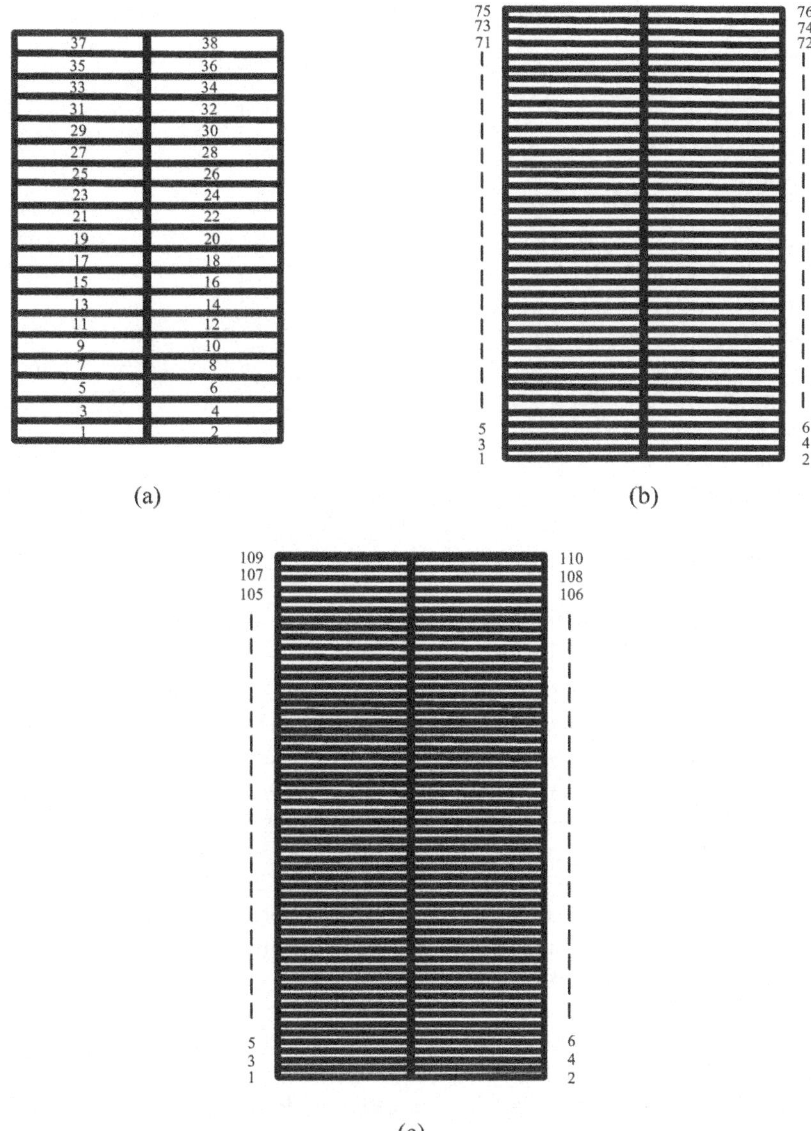

Fig. 6.18 Coil cross section for 13.2 kV, 26.4 kV, and 38.1 kV HG (**a**) Coil cross section for 13.2 kV HG (**b**) Coil cross section for 26.4 kV HG (**c**) Coil cross section for 38.1 kV HG

In order to step up phase voltage to 26.4 kV while maintaining the same turn voltage, that is, 0.023 kV, the number of turns is increased to 76, that is, double the number of turns for the 13.2 kV coils. Therefore, each turn in Fig. 6.18a is split vertically into two turns, and with the same turn arrangement, the coil cross section for the 26.4 kV HG is shown in Fig. 6.18b. The voltage difference between turns is

Table 6.9 HG coil and turn insulations

Description	Initial	Step 1	Step 2	Step 3
Phase voltage (RMS kV)	6.6	13.2	26.4	38.1
Coil voltage (RMS kV)	0.44	0.88	1.76	2.54
Turn voltage (RMS kV)	0.023			
No. of coils per phase	15			
No. of turns per coil	19	38	76	110
Conductor height excluding insulation (mm)	1.64	1.64	0.8279	0.57
Conductor width excluding insulation (mm)	12.2	6.1	6.1	6.1
Conductor insulation (mm)	0.066	0.13	0.13	0.13
Conductor height including insulation (mm)	1.84	1.9	1.08	0.83
Conductor width including insulation (mm)	12.53	6.36	6.36	6.36
Added insulation in the middle of the coil (mm)	–	0.13	0.13	0.13
Groundwall insulation thickness (mm)	2.6			
Coil height (mm)	40.16	41.3	46.24	50.85
Coil width (mm)	17.73	18.05	18.05	18.05
Total coil cross section including insulation (mm)	712.04	745.47	834.63	917.84
Increase in coil height (%)	–	2.83	15.14	26.6
Increase in coil width (%)	–	1.8	1.8	1.8
Increase in coil cross-section area (%)	–	4.7	17.22	28.9

the same as 13.2 kV HG; hence, the same insulation is used as shown in Table 6.9. Compared to the initial values, the coil height, width, and cross-section area for the 26.4 kV HG are increased by 15.14%, 1.8%, and 17.22%, respectively.

To achieve a 38.1 kV RMS phase voltage while maintaining a turn voltage of 0.023 kV, the number of turns per coil needs to increase to 110. Therefore, by keeping the same copper cross-section area, that is, 380.64 mm^2, the same coil copper width and height, that is, 12.2 mm and 31.2 mm, and also by having the same turn arrangements as in Fig. 6.18a, the conductor copper height for the coils in the 38.1 kV machine is 0.57 mm. Therefore, for the same turn arrangements as in 13.2 kV and 26.4 kV HG coils, the conductor insulation thickness stays the same, that is, 0.13 mm. However, due to added number of turns, the coil height, width, and cross-section area are increased by 26.6%, 1.8%, and 28.9%, respectively. The coil cross section for the 38.1 kV HG is shown in Fig. 6.18c while Fig. 6.19 shows stator slot and back-iron cross-sectional view for different designs. Table 6.9 summarizes the winding insulation systems. In order to account for the coil height increase, the slot depth is increased by 26.6%, and to ensure the same machine magnetic performance, the stator back-iron is increased by 26.6% to maintain the same back-iron thickness as in the 6.6 kV HG. Therefore, the stator inner diameter is the same, while the stator outer diameter is increased by 26.6%. As a result, the machine mass is increased by 7.5%.

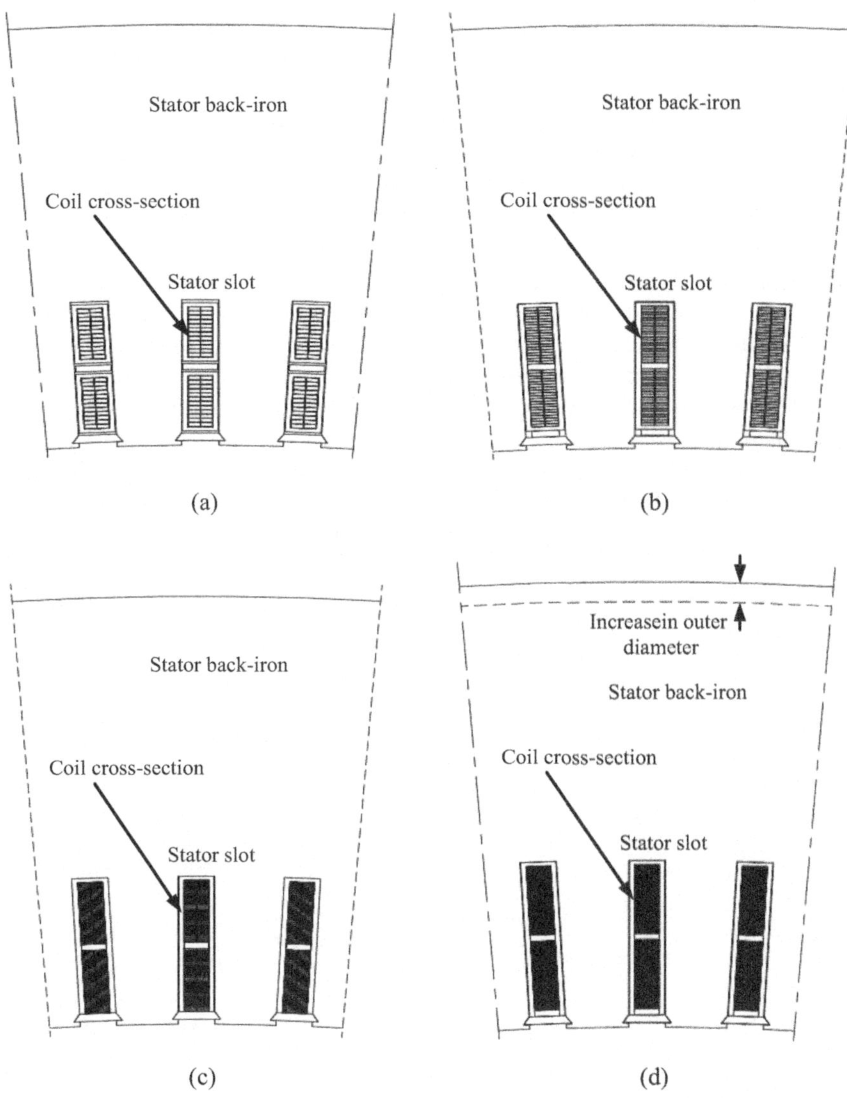

Fig. 6.19 Stator cross-sectional view for different HG winding designs (**a**) 6.35 kV SG (**b**) 13.2 kV HG (**c**) 26.4 kV HG (**d**) 38.1 kV HG

6.6 Summary

The insulation systems for the HV machines are discussed in this chapter. Three different types of windings for the machines are introduced, namely, random-wound coils, form-wound coils, and Roebel bars. The form-wound coils, common between the high-voltage machine manufacturers, is chosen for the HG windings. Insulation systems for the strand, conductor, and coils are considered, and the insulations for the benchmark SG are chosen as the bases of the insulation design for the HG. The importance of the groundwall, semiconductive, and voltage stress grading insulations is illustrated. The winding insulation systems are modeled as capacitor circuits, and the importance of avoiding air pockets is illustrated via an example. The groundwall insulation for the HG is the same as the benchmark SG, while the strand and conductor insulations increase linearly with the voltage. The RMS phase voltage increase from 6.6 kV to the desired 38.1 kV is explained using three steps to illustrate the impact of winding voltage on the machine stator design. In steps 1 and 2, the voltage is doubled, while in step 3, the voltage is increased by a factor of 1.44. The turns in each coil are arranged such that the voltage difference between consecutive coils is minimized, hence minimizing the insulation requirements. It is shown that for the same number of coils per phase and the same turn voltages, the coil height is increased by 26.6% from its original value for the 6.6 kV HG. To account for the coil height increase, the stator slots are made deeper while stator back-iron thickness is kept the same. Therefore, the 38.1 kV HG has an outer diameter of 26.6% longer than that of the 6.6 kV HG which results in 7.5% increase in machine mass.

References

1. [Online] Brazos on-shore wind farm, Texas, US. Available: https://en.wikipedia.org/wiki/Brazos_Wind_Farm
2. [Online] Walney Off-shore Windfarms. Available: http://www.dongenergy.com/
3. [Online] Alstom wind turbines. Available: http://alstomenergy.gepower.com/
4. E. Hau, *Wind Turbines: Fundamentals, Technologies, Application, Economics* (Springer, Berlin, 2006)
5. [Online] Siemens wind turbine SWT-3.6-107. Available: http://www.siemens.com
6. [Online] Enercon E-82 wind turbine. Available: http://www.enercon.de
7. S.V. Bozhko, R. Blasco-Gimenez, R. Li, J.C. Clare, Control of offshore DFIG-based wind farm grid with line-commutated HVDC connection. IEEE Trans. Energy Convers. **22**(1), 71–78 (2007)
8. H. Liu, J. Sun, Voltage stability and control of offshore wind farms with AC collection and HVDC transmission. IEEE J. Emerg. Sel. Top. Power Electron. **2**(4), 1181–1189 (2014)
9. D. Yoon, H. Song, G. Jang, S. Joo, Smart operation of HVDC systems for large penetration of wind energy resources. IEEE Trans. Smart Grid **4**(1), 359–366 (2013)
10. E. Selvaraj, C.P. Sugumaran, M.R. Krishnamoorthi, M.R. Kumar, J. Joshi, S. Ganesan, S. Geethadevi, D. Kumar, A review on fundamentals of HVDC transmission. J. Club Electr. Eng. (JCEE) **1**, 12–17 (2014)
11. R. Feldman, M. Tomasini, E. Amankwah, J.C. Clare, P.W. Wheeler, D.R. Trainer, R.S. Whitehouse, A hybrid modular multilevel voltage source converter for HVDC power transmission. IEEE Trans. Ind. Appl. **49**(4), 1577–1588 (2013)
12. J.A. Baroudi, V. Dinavahi, A.M. Knight, A review of power converter topologies for wind generators. Renew. Energy **32**, 2369–2385 (2007)
13. O. Anaya-Lara, N. Jenkins, J. Ekanayake, P. Cartwright, M. Hughes, *Wind Energy Generation: Modelling and Control* (Wiley, West Sussex, 2009)
14. D. Jovcic, N. Strachan, Offshore wind farm with centralised power conversion and DC interconnection. IET Gener. Transm. Distrib. **3**(6), 586–595 (2009)
15. E. Veilleux, P.W. Lehn, Interconnection of direct-drive wind turbines using a series-connected DC grid. IEEE Trans. Sustain. Energy **5**(1), 139–147 (2014)
16. M. Dahlgren, H. Frank, M. Leijon, F. Owman, L. Walfridsson, "Windformer". ABB review. Tech Rep. **3**, 31–37 (2000)
17. S.M. Muyeen, R. Takahashi, J. Tamura, Operation and control of HVDC-connected offshore wind farm. IEEE Trans. Sustain. Energy **1**(1), 30–37 (2010)

© Springer Nature Switzerland AG 2020   177
O. Beik, A. S. Al-Adsani, *DC Wind Generation Systems*,
https://doi.org/10.1007/978-3-030-39346-5

18. C. Zhan, C. Smith, A. Crane, A. Bullock, D. Grieve, DC transmission and distribution system for a large offshore wind farm. 9th IET international conference AC and DC power transmission, pp. 1–5, 19–21 Oct 2010
19. S.D. Wright, A.L. Rogers, J.F. Manwell, A. Ellis, Transmission options for offshore wind farms in the United States. Proceedings of the American Wind Energy Association annual conference, pp. 1–12, 2002
20. V.G. Agelidis, G.D. Demetriades, N. Flourentzou, Recent advances in high-voltage direct-current power transmission systems. IEEE international conference on industrial technology (ICIT) 2006, pp. 206–213, 15–17 Dec 2006
21. P. Bresesti, W.L. Kling, R.L. Hendriks, R. Vailati, HVDC connection of offshore wind farms to the transmission system. IEEE Trans. Energy Convers. **22**(1), 37–43 (2007)
22. C. Meyer, M. Hoing, A. Peterson, R.W. De Doncker, Control and design of DC grids for offshore wind farms. IEEE Trans. Ind. Appl. **43**(6), 1475–1482 (2007)
23. R. Steigerwald, R. Tompkins, A comparison of high-frequency link schemes for interfacing a DC source to a utility grid, IEEE IAS annual meeting, pp. 759–766, 1982
24. R.W. De Doncker, D.M. Divan, M.H. Kheraluwala, A three-phase soft-switched high-power-density DC/DC converter for high-power applications. IEEE Trans. Ind. Appl. **27**(1), 63–73 (1991)
25. D. Jovcic, Bidirectional, high-power DC transformer. Trans. Power Deliv. **24**(4), 2276–2283 (2009)
26. D. Jovcic, L. Zhang, LCL DC/DC Converter for DC Grids. IEEE Trans. Power Deliv. **28**(4), 2071–2079 (2013)
27. C. Meyer, R.W. De Doncker, Design of a three-phase series resonant converter for offshore DC grids. IEEE industry applications conference, 42nd IAS annual meeting, pp. 216–223, 23–27 Sept 2007
28. L. Max, T. Thiringer, Control method and snubber selection for a 5 MW wind turbine single active bridge DC/DC converter. European conference on power electronics and applications, pp. 1–10, 2–5 Sept 2007
29. W. Chen, A.Q. Huang, C. Li, G. Wang, W. Gu, Analysis and comparison of medium voltage high power DC/DC converters for offshore wind energy systems. IEEE Trans. Power Electron. **28**(4), 2014–2023 (2013)
30. M. Hajian, J. Robinson, D. Jovcic, B. Wu, 30 kW, 200 V/900 V, Thyristor LCL DC/DC converter laboratory prototype design and testing. IEEE Trans. Power Electron. **29**(3), 1094–1102 (2014)
31. N. Denniston, A.M. Massoud, S. Ahmed, P.N. Enjeti, Multiple-module high-gain high-voltage DC–DC transformers for offshore wind energy systems. IEEE Trans. Ind. Electron. **58**(5), 1877–1886 (2011)
32. G. Ortiz, J. Biela, D. Bortis, J.W. Kolar, 1 Megawatt, 20 kHz, isolated, bidirectional 12kV to 1.2kV DC-DC converter for renewable energy applications. International power electronics conference (IPEC), pp. 3212–3219, 21–24 June 2010
33. K.T. Chau, Y.B. Li, J.Z. Jiang, S. Niu, Design and control of a PM brushless hybrid generator for wind power application. IEEE Trans. Magn. **42**(10), 3497–3499 (2006)
34. C. Liu, K.T. Chau, J.Z. Jiang, L. Jian, Design of a new outer-rotor permanent magnet hybrid machine for wind power generation. IEEE Trans. Magn. **44**(6), 1494–1497 (2008)
35. R. Pillai, S. Narayanan, G. Swindale, Benefits and challenges of a grid coupled wound rotor synchronous generator in a wind turbine application. Technical information from Cummins generator technologies, Issue number: WP102
36. I.D. Margaris, N.D. Hatziargyriou, Direct drive synchronous generator wind turbine models for power system studies. 7th Mediterranean conference and exhibition on power generation, transmission, distribution and energy conversion, pp. 1–7, 7–10 Nov 2010
37. S. Achilles, M. Poller, Direct drive synchronous machine models for stability assessment of wind farms. Fourth international workshop on large scale integration of wind power and transmission networks for offshore windfarms, Billund, 2003

38. M.R. Behnke, E. Muljadi, Reduced order dynamic model for variable-speed wind turbine with synchronous generator and full power conversion topology. International conference on future power systems, Nov 2005
39. E. Spooner, P. Gordon, J.R. Bumby, C.D. French, Lightweight ironless-stator PM generators for direct-drive wind turbines. IEE Proc. Electric Power Appl. **152**(1), 17–26 (2005)
40. S. Brisset, D. Vizireanu, P. Brochet, Design and optimization of a nine-phase axial-flux PM synchronous generator with concentrated winding for direct-drive wind turbine. IEEE Trans. Ind. Appl. **44**(3), 707–715 (2008)
41. H. Li, Z. Chen, Overview of different wind generator systems and their comparisons. IET Renew. Power Gener. **2**(2), 123–138 (2008)
42. D.J. Bang, H. Polinder, G. Shrestha, J.A. Ferreira, Review of generator systems for direct-drive wind turbines. Eur. Wind Energy Conf. Exhib., Belgium, Mar 31–Apr 3 2008
43. H. Polinder, F. van der Pijl, G. de Vilder, P.J. Tavner, Comparison of direct-drive and geared generator concepts for wind turbines. IEEE Trans. Energy Convers. **21**(3), 725–733 (2006)
44. M. Liserre, R. Cardenas, M. Molinas, J. Rodriguez, Overview of multi-MW wind turbines and wind parks. IEEE Trans. Ind. Electron. **58**(4), 1081–1095 (2011)
45. [Online] DeWind D8.2 2000 kW wind turbine. Available: http://www.dewindco.com
46. A. Betz, Wind-Energie und ihre Ausnutzung durch Windmühlen (Wind energy and its utilization through windmills), Öko-Buchverlag Kassel, 1982
47. S. Heier, *Grid Integration of Wind Energy Conversion Systems* (Wiley, New York, 1998)
48. Z. Lubosny, *Wind Turbine Operation in Electric Power Systems* (Springer, New York, 2003)
49. O. Wasynczuk, D.T. Man, J.P. Sullivan, Dynamic behavior of a class of wind turbine generators during random wind fluctuations. IEEE Power Eng. Rev. **PER-1**(6), 47–48 (1981)
50. P.M. Anderson, A. Bose, Stability simulation of wind turbine systems. IEEE Trans Power Appar. Syst. **PAS-102**(12), 3791–3795 (1983)
51. Department of Energy, NASA, The Mod-2 Wind Turbine Development Project. Report No. DOE/NASA/20305–5 NASA TM-82681, July 1981
52. B. Amlang, D. Arsurdis, W. Leonhard, W. Vollstedt, K. Wefelmeier, Elektrische Energieversorgung mit Windkraftanlagen Abschlußbericht BMFT-Forschungsvorhaben 032–8265-B, Braunschweig, 1992
53. V. Quaschning, *Understanding Renewable Energy Systems* (Earthscan, London, 2005)
54. G. Bywaters, V. John, J. Lynch, P. Mattila, G. Norton, J. Stowell, M. Salata, O. Labath, A. Chertok, D. Hablanian, Northern Power Systems WindPACT drive train alternative design study report. NREL, Golden, CO, Rep. Number NREL/SR-500-35524, 2004
55. [Online] Available: https://www.motiondynamics.com.au/gearboxes-which-type.html
56. D. Jovcic, Step-up DC–DC converter for megawatt size applications. IET Power Electron. **2**(6), 675–685 (2009)
57. D. Jovcic, High gain DC transformer. U.K. Patent office, PCT Patent application no. GB 0724369.4, Dec 2007
58. R. Datta, V.T. Ranganathan, Variable-speed wind power generation using doubly fed wound rotor induction machine-a comparison with alternative schemes. IEEE Trans. Energy Convers. **17**(3), 414–421 (2002)
59. [Online] Alstom Haliade 150–6MW wind turbines. Available: http://alstomenergy.gepower.com/
60. T. Surinkaew, I. Ngamroo, Coordinated robust control of DFIG wind turbine and PSS for stabilization of power oscillations considering system uncertainties. IEEE Trans Sustain. Energy **5**(3), 823–833 (2014)
61. A.D. Hansena, P. Sørensena, F. Iovb, F. Blaabjerg, Centralised power control of wind farm with doubly fed induction generators. Renew. Energy **31**(2), 935–951 (2006)
62. L.M. Fernandez, C.A. Garcia, F. Juradob, Comparative study on the performance of control systems for doubly fed induction generator (DGIG) wind turbines operating with power regulation. Energy **33**(2), 1438–1452 (2008)
63. W. Lu, B.T. Ooi, Multi-terminal LVDC system for optimal acquisition of power in wind-farm using induction generators. IEEE Trans. Power Electron. **17**(4), 558–563 (2002)

64. W. Lu, B.T. Ooi, Optimal acquisition and aggregation of offshore wind power by multiterminal voltage-source HVDC. IEEE Trans. Power Deliv. **18**(1), 201–206 (2003)
65. N. Schofield, Electric machine design and operation. Ph.D. and M.Sc. lecture notes, Department of Electrical and Computer Engineering, McMaster University, Hamilton, Ontario, Canada, 2015
66. S.J. Sugimoto, Current status and recent topics of rare-earth permanent magnets. J. Phys. D. Appl. Phys. **44**(6), 064001 (2011)
67. M. Humphries, Rare earth elements: the global supply chain. CRS Report for Congress, Report No. R41347, Congressional Research Service, 2013
68. C. Hurst, *China's Rare Earth Elements Industry: What Can the West Learn?* (Institute for the Analysis of Global Security (IAGS), Washington, DC, 2010)
69. [Online] Arnold Magnetics permanent magnets datasheet. Available: http://www.arnoldmagnetics.com/en-us/
70. D. Ishak, A.Q. Zhu, D. Howe, Eddy-current loss in the rotor magnets of permanent-magnet brushless machines having a fractional number of slots per pole. IEEE Trans. Magn. **41**(9), 2462–2469 (2005)
71. A. Boglietti, A. Cavagnino, M. Lazzari, M. Pastorelli, Two simplified methods for the iron losses prediction in soft magnetic materials supplied by PWM inverter. IEEE international on electric machines and drives conference (IEMDC 2001), pp. 391–395, 2001
72. A. Krings, J. Soulard, Overview and comparison of Iron loss models for electrical machines. J. Electr. Eng. **10**(3), 162–169 (2010)
73. W.A. Roshen, A practical, accurate and very general Core loss model for nonsinusoidal waveforms. IEEE Trans. Power Electron. **22**, 30–40 (2007)
74. C.P. Steinmetz, On the law of hysteresis. Proc. AIEEE **72**, 197–221 (1884)
75. Lamination Steel Third Edition CD-ROM, Electric Motor Education and Research Foundation, ISBN 0–9714391–3-3, Second Printing 2009
76. B.J. Moore, R.H. Rehder, R.E. Draper, Utilizing reduced build concepts in the development of insulation systems for large motors, in *Proceedings of IEEE Electrical Insulation Conference*, (Cincinnati, 1999, Oct), pp. 347–352
77. G.C. Stone, E.A. Boulter, I. Culbert, H. Dhirani, *Electrical Insulation for Rotating Machines* (IEEE Press, New York, 2004), p. 5
78. [Online] Partzsch Group. Available: https://en.partzsch.de/roebel-bars
79. M. Leijon, ABB Powerformer. Report No: ABB Review 2/1998
80. X. Yu, W. Wu, W. Pan, S. Han, L. Wang, J. Wei, L. Liu, S. Du, Z. Zhou, A. Foussat, P. Libeyre, Development of insulation technology with vacuum-pressure-impregnation (VPI) for ITER correction coil. IEEE Trans. Appl. Supercond. **22**(3), 7700504–7700504 (2012)
81. M.K.W. Stranges, D.A. Snopek, A.K. Younsi, J.H. Dymond, Effect of surge testing on unimpregnated ground insulation of VPI stator coils. IEEE Trans. Ind. Appl. **38**(5), 1460–1465 (2002)
82. W. Grubelnik, C. Stiefmaier, Un-impregnated vpi tape testing and effects on dielectric performance of VPI insulation systems. Electrical insulation conference (EIC), pp. 359–362, 8–11 June 2014
83. A. Nakayama, H. Haga, M. Muraoka, Development of generator stator coil with 22kV global VPI insulation. Conference on electrical insulation and dielectric phenomena (CEIDP '04), pp. 224–227, 17–20 Oct 2004
84. F.P. Espino-Cortes, E.A. Cherney, S. Jayaram, Effectiveness of stress grading coatings on form wound stator coil groundwall insulation under fast rise time pulse voltages. IEEE Trans. Energy Convers. **20**(4), 844–851 (2005)
85. D. J. Conley, N. Frost, Fundamentals of semi-conductive systems for high voltage stress grading. Electrical insulation conference and electrical manufacturing expo, pp. 89–92, 26–26 Oct 2005
86. Private conversation with Dr. Yali (Natalie) Feng, HV insulation expert at Sulzer Dowding & Mills, Birmingham, UK, 2014

References for Further Studies

W.-L. Chen, Y.-Y. Hsu, Controller design for an induction generator driven by a variable-speed wind turbine. IEEE Trans. Energy Convers. **21**(3), 625–635 (2006)

W. Yang, P.J. Tavner, M.R. Wilkinson, Condition monitoring and fault diagnosis of a wind turbine synchronous generator drive train. IET Renew. Power Gencr. **3**(1), 1–11 (2009)

S. Muller, M. Deicke, R.W. De Doncker, Doubly fed induction generator systems for wind turbines. IEEE Ind. Appl. Mag. **8**(3), 26–33 (2002)

K. Tan, S. Islam, Optimum control strategies in energy conversion of PMSG wind turbine system without mechanical sensors. IEEE Trans. Energy Convers. **19**(2), 392–399 (2004)

[Online] Article. Available: http://cleantechnica.com/2014/05/07/wind-power/

[Online] Direct-Connected Induction (Asynchronous) Generator. Available: http://www.uwig.org/

Z. Tan, X. Song, W. Cao, Z. Liu, Y. Tong, DFIG machine design for maximizing power output based on surrogate optimization algorithm. IEEE Trans. Energy Convers **30**(3), 1154–1162 (2015)

V. Delli Colli, F. Marignetti, C. Attaianese, Analytical and multiphysics approach to the optimal design of a 10-MW DFIG for direct-drive wind turbines. IEEE Trans. Ind. Electron. **59**(7), 2791–2799 (2012)

M. Fazeli, S.V Bozhko, G.M. Asher, L. Yao, Voltage and frequency control of offshore DFIG-based wind farms with line commutated HVDC connection. 4th IET power electronics, machines and drives (PEMD), pp. 335–339, 2–4 April, 2008

A. Boumassata, D. Kerdoun, Direct powers control of DFIG through direct converter and sliding mode control for WECS. 3rd international conference on control, engineering & information technology (CEIT), May 2015

S. Ghosh, Closed loop control of a DFIG based wind power system IEEE innovative smart grid technologies – Asia (ISGT Asia), 10–13 Nov 2013

W. Srirattanawichaikul, Y. Kumsuwan, S. Premrudeepreechacharn, B. Wu, A vector control of a grid-connected 3L-NPC-VSC with DFIG drives. International conference on electrical engineering/electronics computer telecommunications and information technology (ECTI-CON), pp. 828–832, 19–21 May 2010

Y. Lei, A. Mullane, G. Lightbody, R. Yacamini, Modeling of the wind turbine with a doubly fed induction generator for grid integration studies. IEEE Trans. Energy Convers. **21**(1), 257–264 (2006)

G.L. Johnson, *Wind Energy Systems* (Prentice Hall, Englewood Cliffs, 1985)

R. Gasch, J. Twele, *Wind Power Plants: Fundamentals, Design, Construction and Operation* (Solarpraxis AG, Berlin, 2002)

Y. Xia, K.H. Ahmed, B.W. Williams, Wind turbine power coefficient analysis of a new maximum power point tracking technique. IEEE Trans. Ind. Electron. **60**(3), 1122–1132 (2013)

MATLAB software Powersim Simulink help

M. Chinchilla, S. Arnaltes, J.C. Burgos, Control of permanent-magnet generators applied to variable-speed wind-energy systems connected to the grid. IEEE Trans. Energy Convers. **21**(1), 130–135 (2006)

N. Milivojevic, N. Schofield, I. Stamenkovic, Y. Gurkaynak, Field weakening control of PM generator used for small wind turbine application. IET conference on renewable power generation (RPG 2011), 6–8 Sept 2011

Department of Defense, The military handbook for reliability prediction of electronic equipment. Tech. Rep. MIL-HDBK-217F-2, 1995

[Online] Xian Electric Transformer, Transformer & reactor 10kV~500kV, Available: http://www.xianelectric.com

[Online] NKT Cables, Product catalogue Australia medium and high voltage cables, Available: http://www.nktcables.com

[Online] Prysmian Cables, High voltage cables, Available: http://prysmiangroup.com/

N. Mohan, T. Undeland, W. Robbins, *Power Electronics: Converters, Applications, and Design*, 3rd edn. (Wiley, New York, 2003)

J.D. Glover, S.M. Sarma, T.J. Overbye, *Power System Analysis and Design*, 5th edn. (Cengage Learning, Stamford, 2012)

W.L. Kling, P. Bresesti, I. Valade, D. Canever, R.L. Hendriks, Transmission systems for offshore wind farms in the Netherlands. International conference & exhibition "offshore wind", Copenhagen, Denmark, Oct 2005

H. Li, Z. Chen, Design optimization and evaluation of different wind generator systems. International conference on electrical machines and systems (ICEMS), Oct 2008

H. Li, Z. Chen, H. Polinder, Optimization of Multibrid permanent-magnet wind generator systems. IEEE Trans. Energy Convers. **24**(1), 82–92 (2009)

Y.-S. Xue, L. Han, H. Li, L.-D. Xie, Optimal design and comparison of different PM synchronous generator systems for wind turbines. International conference on electrical machines and systems (ICEMS), Oct 2008

D. Kowal, L. Dupre, P. Sergeant, L. Vandenbossche, M. De Wulf, Influence of the electrical steel grade on the performance of the direct-drive and single stage gearbox permanent-magnet machine for wind energy generation, based on an analytical model. IEEE Trans. Magn. **47** (12), 4781–4790 (2011)

R.P. Severns, G. Bloom, *Modern DC-to-DC Switchmode Power Converter Circuits* (Van Nostrand Reinhold Co., New York, 1985)

L. Max, S. Lundberg, System efficiency of a DC/DC converter-based wind farm. J. Wind Energy **11**, 109–120 (2008)

Y. Zhou, D.E. Macpherson, W. Blewitt, D. Jovcic, Comparison of DC-DC converter topologies for offshore wind-farm application. IET international conference on power electronics, machines and drives (PEMD), 2012

[Online] Vestas wind turbines. Available: http://www.vestas.com

[Online] Siemens wind turbine. Available: http://www.offshorewind.biz/

[Online] Liquid filled transformers. Available: http://new.abb.com/products/transformers

[Online] ABB Converter, Low voltage wind turbine converter ACS880, 800 kW – 8 MW, Document No. 3AXD10000303002 02.2014. Available: www.abb.com/converters-inverters

[Online] Available: http://hvdiode.en.alibaba.com/

R. Pena, J.C. Clare, G.M. Asher, Doubly fed induction generator using back-to-back PWM converters and its application to variable-speed wind-energy generation. IEE Proc. Electric Power Appl. **143**(3), 231–241 (1996)

[Online] SG in a wind turbine. Available: http://www.alternative-energy-tutorials.com/wind-energy/synchronous-generator.html

J. Hansson, Analysis and control of a hybrid vehicle powered by a free-piston energy converter. PhD Thesis, Royal Institute of Technology (KTH), Stockholm, Sweden 2006

S. Jordan, C.D. Manolopoulos, J. Apsley, Winding configurations for five-phase synchronous generators with diode rectifiers. IEEE Trans. Ind. Electron. **63**(1), 517–525 (2016)

A. S. Al-Adsani, Hybrid permanent magnet machines for electric vehicles. PhD Thesis, The University of Manchester, Manchester, UK, 2006

O. Beik, N. Schofield, A brushless exciter design for a hybrid permanent magnet generator applied to series hybrid electric vehicles. The 7th IET international conference on power electronics, machines and drives PEMD'14, Manchester, UK, April 2014

E. Levi, M. Jones, D. Dujic, *Electric Multiphase Motor Drives: Modeling and Control* (CRC Press LLC, Berlin, 2015)

I. Boldea, S.A. Nasar, *Electric Machine Dynamics* (Macmillan, Oxford, UK, 1986)

V. Del Toro, *Electric Machines and Power Systems* (Prentice-Hall, Englewood Cliff, 1985)

J.I. Grainger, W. Stevenson Jr., *Power System Analysis* (McGraw-Hill, New York, 1994)

J.L. Rodriguez-Amenedo, S. Arnalte, J.C. Burgos, Automatic generation control of a wind farm with variable speed wind turbines. IEEE Trans. Energy Convers. **17**(2), 279–284 (2002)

J.F. Gieras, *Advancements in Electric Machines* (Springer, Dordrecht, 2010)

L. Li, X. Huang, B. Kao, B. Yan, Research of core loss of permanent magnet synchronous motor (PMSM) in AC servo system, International conference on electrical machines and systems (ICEMS 2008), pp. 602–607, 17–20 Oct 2008

G. Bertotti, A. Boglietti, M. Chiampi, D. Chiarabaglio, F. Fiorillo, M. Lazzari, An improved estimation of iron losses in rotating electrical machines. IEEE Trans. Magn. **27**(6), 5007–5009 (1991)

G.R. Slemon, X. Liu, Core losses in permanent magnet motors. IEEE Trans. Magn. **26**(5), 1653–1655 (1990)

K. Atallah, Z.Q. Zhu, D. Howe, An improved method for predicting iron losses in brushless permanent magnet DC drives. IEEE Trans. Magn. **28**, 2997–2999 (1992)

I. Mayergoyz, Mathematical models of hysteresis. IEEE Trans. Magn. **22**, 603–608 (1986)

L.R. Dupre, O. Bottauscio, M. Chiampi, M. Repetto, J.A.A. Melkebeek, Modeling of electromagnetic phenomena in soft magnetic materials under unidirectional time periodic flux excitations. IEEE Trans. Magn. **35**, 4171–4184 (1999)

G. Bertotti, General properties of power losses in soft ferromagnetic materials. IEEE Trans. Magn. **24**(1), 621–630 (1988)

F.T. Emery, D.C. Johnson, Voltage grading model for high voltage electric generator stator coil end turn regions. Conference on electrical insulation and dielectric phenomena, pp. 589–592, 25–28 Oct 1998

A. Roberts, Stress grading for high voltage motor and generator coils. IEEE Electr. Insul. Mag. **11**(4), 26–31 (1995)

M. Fujita, Y. Kabata, T. Tokumasu, K. Nagakura, M. Kakiuchi, S. Nagano, Circulating currents in Stator coils of large turbine generators and loss reduction. IEEE Trans. Ind. Appl. **45**(2), 685–693 (2009)

Y. Liang, X. Bian, H. Yu, L. Wu, B. Wang, Analytical algorithm for strand end leakage reactance of transposition Bar in AC machine. IEEE Trans. Energy Convers. **30**(2), 533–540 (2015)

Index

© Springer Nature Switzerland AG 2020
O. Beik, A. S. Al-Adsani, *DC Wind Generation Systems*,
https://doi.org/10.1007/978-3-030-39346-5

The manufacturer's authorised representative in the EU is Springer
Nature Customer Service Centre GmbH, Europaplatz 3, 69115 Heidelberg,
Germany. If you have any concerns regarding our products, please
contact ProductSafety@springernature.com

Printed and bound by CPI Group (UK) Ltd, Croydon, CR0 4YY
28/11/2025
02007692-0004